COLLABORATIVE DESIGN MANAGEMENT

The design process has always been central to construction, but recent years have seen its significance increase, and the ways of approaching it multiply. To an increasing degree, other stakeholders such as contractors have input at the design stage, and the designer's role includes tasks that were traditionally the realm of other professions.

This presents challenges as well as opportunities, and both are introduced, discussed, and analysed in *Collaborative Design Management*. Case studies from the likes of Arup, Buro Happold, VINCI Construction UK Limited, and the CIOB show how technologies (BIM, podcasting), innovative working (information management, collaboration), and the evolution of roles (the designer–contractor interface, environmental compliance) have changed design management as a process.

Starting from a basic level, the reader is introduced to the key themes and background to the design management role, including definitions of the responsibilities now commonly involved, and the strategic importance of design. Influential technologies currently in use are evaluated, and the importance they are likely to have in the future is explored.

This combination of case studies from leading practitioners, clear explanations of design management roles and activities, and an exploration of how to successfully achieve collaborative design management makes this a highly topical and uniquely valuable book. This is essential reading for professionals and students of all levels interested in construction design management, from all AEC backgrounds.

Stephen Emmitt is Professor of Architectural Technology in the School of Civil and Building Engineering at Loughborough University, UK. He formerly held the Hoffmann Chair of Innovation and Management at the Technical University of Denmark and is currently Visiting Professor of Construction Innovation at Halmstad University, Sweden.

Kirti Ruikar is a Senior Lecturer in Architectural Engineering in the School of Civil and Building Engineering at Loughborough University, UK. She is currently Joint Coordinator of CIB Task Group, TG83 on 'E-business in Construction' and an Associate Editor of the Journal of IT in Construction (ITcon.org).

COLLABORATIVE DESIGN MANAGEMENT

Stephen Emmitt and Kirti Ruikar

Routledge
Taylor & Francis Group

LONDON AND NEW YORK

First published 2013
by Routledge
2 Park Square, Milton Park, Abingdon, Oxon, OX14 4RN

Simultaneously published in the USA and Canada
by Routledge
711 Third Avenue, New York, NY 10017

Routledge is an imprint of the Taylor & Francis Group, an informa business

British Library Cataloguing in Publication Data
A catalogue record for this book is available from the British Library

Library of Congress Cataloging-in-Publication Data
Emmitt, Stephen.
Collaborative design management / Stephen Emmitt and Kirti Ruikar.
p. cm
Includes bibliographical references and index.
1. Architects and builders. 2. Building–Superintendence. I. Ruikar, Kirti. II. Title.
NA2543.B89E46 2013
690–dc23
2012037485

ISBN13: 978-0-415-62074-1 (hbk)
ISBN13: 978-0-415-62075-8 (pbk)
ISBN13: 978-0-203-81912-8 (ebk)

Typeset in Frutiger
by FiSH Books Ltd, Enfield

Printed and bound in Great Britain by
TJ International Ltd, Padstow, Cornwall

CONTENTS

ACKNOWLEDGEMENTS

We would like to take this opportunity to extend our gratitude to our industry friends and colleagues, whose valuable case study contributions have added the much needed practical context to our book. Their experiences are invaluable in providing a deeper insight into the challenges facing the construction sector; and also in providing examples of approaches devised to address challenges and hence drive innovation. Our contributors are:

Paula Bleanch, Northumbria University
Rachel Bowen-Price, Morgan Sindall
Dan Clipsom, Arup
Philip Davies, VINCI Construction UK Limited
Gerard Daws, NBS Schumann Smith
John Eynon, Chair CIOB Design Management Group/Open Water Consulting
Michael Graham, UKValueManagement
Matt Griffiths, Thomas Vale Construction
Paul Hill, Arup
Tanya Ross, Buro Happold
Darshan Ruikar, Arup
Chiara Tuffanelli, Arup
Tom Warden, Tesco PLC (Property)
Paul Wilkinson, pwcom.co.uk Ltd

Stephen Emmitt and Kirti Ruikar
Loughborough University

INTRODUCTION

Since the turn of the century there has been a rapid increase in the number of design managers employed by contracting organisations in the UK. This trend has also been witnessed in many other countries around the globe as contractors seek to deliver better value to their customers and also respond to the challenge of constructing environmentally sustainable buildings. In responding to these and associated drivers, such as more stringent legislation, rapidly changing technologies, evolving procurement systems, and greater responsibility for design quality, many contracting organisations have recognised the value of design. In responding to a changing marketplace, contracting organisations have started to appreciate the impact of effective design management throughout the entire lifecycle of a project, resulting in increased efficiency and competitiveness, and hence better value for their patrons. In turn, this has brought about increased knowledge and understanding of the strategic power that design management brings to their businesses; with design management integral to the management of the project portfolio and the management of the organisation.

The collaborative and integrative nature of the design management role appears to fit very well in many contracting organisations as the constructors take on greater responsibility for individual design packages and with it the overall quality of building design. Working alongside project managers and construction managers, it is the construction design managers who take overall accountability for design quality; coordinating design information, reviewing designs and managing design changes. Addressing design value helps to promote leaner processes and realise greener buildings, resulting in less waste and better value for clients and building users. Managing design value also helps contracting organisations to improve their effectiveness, resulting in higher levels of profitability. Fundamental to unlocking the potential of good design is the ability to work collaboratively with a large number of disciplines and trades, and with a diverse range of organisations and individuals, in a temporal project coalition.

Desire for closer integration and better collaboration between the participants in construction projects has been helped and facilitated by developments in information technologies. Information communication technologies (ICTs) such as project extranets, intranets and mobile technologies allow project participants to communicate and share information much more easily than was previously possible. Advances in software design have resulted in the development and use of building information modelling and building information management (BIM) packages. ICTs allow the development of building designs in a collaborative environment, helping to resolve common challenges such as the coordination and flow of information as well as identifying clash detection long before it could pose a problem on the construction site. Building information models allow the building to be constructed in a virtual environment, allowing designers and constructors to model the construction process to find the most effective, and safe, process for assembly and disassembly before moving into the physical act of building. Thus the building is 'built' twice, first virtually and then physically. Digital technologies are making a significant impact on the manner in which buildings are designed, constructed and maintained. Developments in ICTs and BIM also impacts on the day-to-day tasks of the construction design manager; an area that will quickly evolve as BIM is taken up more widely over the coming years.

Many of the individuals currently working as construction design managers have previously held positions as architects, architectural technologists, engineers and project managers, and now find themselves, through choice or circumstance, in a new and rapidly evolving role. Terms such as 'design manager', 'design integrator' or 'design coordinator' have become commonplace, although the exact nature of the role has not always been well defined or understood. Part of the confusion appears to stem from the wide interpretation and application of the role in industry; partly from the paucity of education and training programmes for aspiring design managers; and partly from the way in which the terms design management and design manager are used in separate fields of literature. Given that design management is an emergent discipline in the construction sector, a lack of clarity is to be expected as industry refines the role and academics try to make sense of a rapidly changing landscape.

Background to an evolving discipline

Although the construction design management role has been eluded to in literature going back to the 1960s, the emergence of the role in the UK construction sector appears to have occurred in the early 1990s as contractors adopted new procurement routes, and in doing so took greater responsibility for managing design. Contractors had previously employed 'resident architects' and 'resident engineers' to help review and coordinate design information and deal with design changes. As design and build started to increase in popularity from the 1980s onwards, the major contractors started to appreciate the need to manage design, rather than just coordinate design information. A report published by Reading University *The Successful Management of Design* (Gray *et al.*, 1994), later developed into a book by Gray and Hughes (2001), helped to further emphasise the importance of design to the contracting fraternity. These developments coincided with increasingly complex construction projects, the growing need to coordinate information better and the desire of the specialists involved in construction projects to collaborate more effectively. Reports published in the 1990s (e.g. Latham, 1994; Egan, 1998, 2002) questioned the ability of the construction sector as a whole to deliver value to the construction client (the customer), and by implication, society. Reflecting the sentiments of earlier publications – Phillips (1950) had emphasised the need for better collaboration, management and coordination; Emerson (1962) the need for better communication – these reports urged a change in attitude and culture – urging, amongst other things, more collaboration, closer integration of work, and leaner working methods. This was coincident with clients starting to seek better value from their projects and in doing so demanding more for less financial outlay, better (environmental) performance, and higher quality levels: all areas in which design has a significant role to play. Combined with research findings and associated articles in the professional trade press, the Latham and Egan reports brought about pressure to change, with the adoption of new forms of contract, new ways of working, and with it shifting roles and responsibilities. One of these changes was further growth of contractor led procurement and with it the development of the design manager role in the UK (Gray and Hughes, 2001; Bibby, 2003) and internationally (e.g. Grilo *et al.*, 2007; Beim and Jensen, 2007). As the number of design managers increased, the job function started to be recognised as something different from a project manager or a construction manager. In 2007, the construction design manager role was formally recognised by the Chartered Institute of Builders (CIOB). The majority of the large and medium-sized contractors now have some form of in-house design management or design coordination team. Similarly, the management consultants are offering design management services to their clients. The design management function has also spread to many large client organisations with significant property portfolios, such as the food retail sector, with design managers instrumental in site identification and design development.

Early application of design management by contractors appears to have been confined to the construction site, with construction design managers being responsible for coordinating information, addressing buildability issues, dealing with requests for information, and managing the impact and cost of design changes. In this role the construction design manager works closely with the project managers and the site-based construction managers. More recently the remit has expanded to include responsibilities for the management of design during the pre-contract phase, ranging from client briefing and managing various approval processes, through to the detailed design phase. Both functions are to be found on new build and refurbishment projects, illustrating the rapid evolution of the discipline into two distinct roles (and requiring different skills sets); a point explored in more detail in Chapter 4. In both jobs, the construction design manager remit will cover the management of architectural and structural design as well as mechanical and electrical services design. Including all design stages has been instrumental in helping some contracting organisations to offer a 'one stop shop' for their clients, and in doing so promoting the benefits of integrated design as a vehicle to improve the efficiency of projects and the performance of buildings.

Expansion of the design managers' remit has allowed contractors to address buildability issues earlier in the process and better integrate technologies to improve the environmental performance of buildings. Architects have long recognised the benefits of getting the design correct and the information complete before construction work starts. Unfortunately, this message is not always communicated to the client and the contractor, with the result that there is considerable pressure to start the physical act of building before decisions and information is complete. This results in requests for information, requests for design changes and additional (one could argue unnecessary) work for the constructors. By managing the design in such a way that design decisions and information are complete it is possible to start on site with increased certainty and less risk. Constructors are also better able to coordinate the design and construction work, resulting in more efficient and less wasteful work. Thus collaborative design management should be viewed as a value-adding activity.

The constructors' thirst for construction design managers has led to innovations in education. An undergraduate programme designed entirely in response to the needs of contractors was started at Loughborough University in 2001. The BSc (Hons) Architectural Engineering and Design Management (AEDM) programme came about following interaction with major contractors and their request for a 'new' professional who was educated in design management. The result was an undergraduate programme that emphasises the value of design, integration and collaboration, which is underpinned by a sustainable design philosophy. The undergraduate programme at Loughborough University has proved to be very popular since its launch, with design management graduates joining major contracting organisations in the UK and overseas to work on complex and prestigious projects. Before the launch of the AEDM programme there were very few opportunities for individuals to gain education in construction design management. Now, a small number of institutions offer similar programmes at undergraduate level and increasingly at postgraduate level. Given that the numbers of design management graduates are still small in comparison to the established disciplines it is not surprising that confusion and misunderstanding about the discipline remain.

In many respects, it is the large and medium sized contractors that are setting the agenda for the construction design management field, with most research and teaching following industry application. However, many of the smaller contracting organisations are addressing design management in some form, although the role is not always explicit. Reviewing the small body of literature reveals inconsistencies in philosophy, theory and use of terminology; which are characteristics of an emergent discipline. These characteristics are mirrored in industry, with wide variations in interpretation, application and understanding of the design manager role by contracting organisations. Tzortzopoulos and Cooper (2007) have argued for improved clarity of the role so that different stakeholders can apply appropriate tools and methods to establish the most effective processes, and hence generate best value. However, there is still very little guidance currently available to students or practitioners studying and working with the construction design management field; an observation that led to this book.

Terms and responsibilities

Given the lack of clarity, it is necessary to define the term design management in the context of this book. To distinguish the role in construction from other fields, the term 'construction design manager' is used to signify an individual working as a design manager for a contracting organisation. Construction design managers will collaborate with many others during the life of a project, such as bid managers, estimators, contracts managers, construction managers, architects, engineers, manufacturers and specialist sub-contractors. The construction design manager's day-to-day tasks will overlap with others, most notably the project managers and the construction managers. The responsibilities of each (in simple terms) are:

- Project management. In the majority of contracting organisations, it is the project manager who has ultimate responsibility for the project. Typical performance criteria relate to completion of the project on time, to budget and to specified quality. Although design should feature strongly in the 'quality' criteria, it has not always been a major concern of project managers, the majority of whom are not educated in design, often leading to criticism of the design quality of finished projects.
- Design management. The construction design manager is responsible for all aspects of design, be it pre-contract or post-contract. Although the role encompasses many project management skills, a passion for design quality makes the role unique. It is the design manager who provides leadership in design.
- Construction management. Construction managers are concerned with realising the design safely and efficiently on the construction site. Their concern is with the effective management of resources, such as plant, people and materials. Their task is to translate the design, codified in drawings and specifications, into a physical artefact, thus their attention is on the accuracy and completeness of the information provided to them, not on the quality of design *per se*. Requests for design information and requests for design changes will be channelled through the construction design manager.

What do design managers manage?

Designers are paid to design; engineers to engineer. It is the design manager who is employed to oversee (manage) all design activity to ensure a consistent and coordinated approach to the project. This relieves the designers and engineers of unnecessary administrative and managerial burdens, so that they can concentrate on what they do best. If we ask ourselves what design managers manage, we come to a relatively simple answer, people and information:

- People: design as an activity involves interaction in the act of producing designs. This is carried out primarily within professional design offices and increasingly collaboratively through the use of collaborative information technologies. The output of the design process is design information.

- Information: design involves interaction to create, review and coordinate a vast quantity of information to prevent errors and to ensure accuracy. This information must be translated by constructors into a physical artefact. One of the design manager's most important tasks is to review information provided by architects, engineers and specialists to ensure that the building can be constructed safely and efficiently.

The process of designing and constructing a building comprises various 'threads' of inter-linking information, which vary in complexity and which evolve into new forms of information as the project progresses. The process of transformation from one information state to another is the result of a decision making process, driven by knowledge and information (Hicks et al., 2002). Today's design and construction offices are awash with information systems that perform complex modelling tasks, which help to simulate effective design and build processes. Contained within these systems are complex threads of information, which collectively weave together the fabric of the building project. These threads contain information about the various building functions, such as its aesthetics, circulation, zones, services, safety, lifecycle costs, environment, operations and maintenance, etc. Thus, the challenge for the design manager is two-fold: first, a need to understand the project context and its information needs and second, to understand the systems within which the project knowledge is contained.

The recent developments in computer mediated 'single model' collaborative tools (e.g. BIM), combined with the UK government's mandate to use BIM on construction projects, has set a new trend for designing and developing building projects. This has added another dimension to the skills requirement: design managers need to possess knowledge of emerging technologies so that the processes of creating, manipulating, storing, sharing and using information resources are adequately supported. Managing these aspects requires impeccable coordination of information and processes, cooperation between team members and collaboration within the project portfolio. Thus design managers must stay informed of technological developments and renew their technical know-how so that the various design and construction processes are coordinated.

To be effective in the role, the construction design manager will also need to understand how designers and engineers work, and be able to communicate effectively across a broad spectrum of organisations and levels. This calls for a collaborative approach, excellent interpersonal ('soft') skills, and the ability to make informed decisions on a strategic and operational level.

- Strategic decision making. Strategic decisions are concerned with the long-term direction of a project or organisation. It is the strategic decisions that set the agenda for the effectiveness and profitability of each project. At a strategic level, the design manager will be working closely with the project manager to ensure project deliverables are met.
- Operational decision making. Operational decisions concern day-to-day problem solving in the workplace. Operational decisions are about getting

tasks completed and are concerned with the flow of resources (information, people and materials) and the adherence to processes. At the operational level, the design manager will be liaising with a wide range of designers and forming the interface between the designers and the constructors.

In the majority of contracting organisations a hierarchy is used for reporting and decision making, with construction design managers and construction managers reporting directly to the project manager.

Scope of the book

Motivation for writing this book comes from the authors' direct experience of teaching aspects of collaborative design management on the Architecture Engineering and Design Management (AEDM) BSc (Hons) at Loughborough University. This, together with feedback from graduate construction design managers working for major contracting organisations, highlighted the need for a balanced, clearly structured and informative book. A small number of books are available for architects and other design professionals, but there is very little that addresses design management from the contractors' perspective (Gray and Hughes, 2001 being a notable exception). Our aim is to provide an introduction to the field and also to provide clear guidance for readers studying to be, and working as, construction design managers. A number of case studies from industry are included to illustrate specific issues and recent developments in the design management role.

Chapters develop a logical narrative, with the first three chapters (Chapters 1–3) exploring collaborative working, the characteristics of design and design management respectively. In Chapter 4 we take a detailed look at the role of the construction design manager. This is followed by a chapter that deals with how we discuss design issues and how we review designs. In Chapters 6 and 7 we deal with collaborative technologies and building information modelling (BIM). This sets the scene for Chapter 8, in which strategies for collaborative working are addressed. In the final chapter (Chapter 9) we discuss the future role of the design manager in a digital, collaborative, arena.

We are conscious that the (construction) design management role is interpreted and applied differently among the contracting community in the UK and also internationally. Similarly, the interpretation and degree of collaborative working varies within organisations, within projects and within different countries. As such, our book should be seen as a primer, a source of information and inspiration, rather than one that tries to provide answers to complex and context specific challenges. We also hope the contents go some way to inform and inspire future generations of design managers in the AEC sector.

1

COLLABORATIVE WORKING

There is nothing particularly new about the concept of collaboration in construction, with master builders collaborating with workers and clients to realise their objectives long before the establishment of the building professions. Similarly, the concepts of collaboration and teamwork have featured within the literature for a long time. What have changed are the technologies available to designers and constructors as the digital revolution gathers pace. Now it is possible to collaborate in real time as easily with project collaborators geographically located the other side of the world as it is with those physically located in the same place. This, in theory at least, makes the task of designing a building much easier, although we need to remember that design is a social task and that the individuals contributing to the design process need to be able to work together effectively, i.e. they need to share some common values (Thyssen, 2011). In this chapter we introduce some of the main issues that underpin collaborative working.

Background

It was during the industrial revolution that some architects and designers made a strong argument for craft and collaboration. Phillip Webb, the pioneer of the Arts and Crafts movement and architects such as John Ruskin and William Morris promoted collaboration between architects and tradesmen, while Corbusier argued for closer collaboration between architects and engineers. The Bauhaus movement also pursued a collaborative ideal in its earlier years. In particular, Walter Gropius was devoted to the notion of collaboration between the fields of design and architecture eventually setting up the *Architects Collaborative* in 1945 (Gropius and Harkness, 1966).

In the UK, the design of buildings does not lie exclusively within the domain of the professional designers, such as the architects and design engineers. Although it may be desirable, one might argue essential, to engage the services of an architectural practice to design a building, many new and refurbishment projects are designed by others with varying degrees of design ability, or are 'borrowed' from standard building layouts. In the best examples, design is very much a collaborative activity, with project participants working together to try and realise the best possible solution for the client given the restraints of time, resources and budget.

A small number of government publications has been highly influential in bringing about greater awareness and subsequent attention to the importance of effective working. Two publications, *Trust and Money* (Latham, 1993) and *Constructing the Team* (Latham, 1994), were significant in raising awareness of teamwork, collaborative working and project partnering. These fundamental tenets of construction projects were subsequently reinforced in *Rethinking Construction* (Egan, 1998) and *Accelerating Change* (Egan, 2002). Combined, the Latham and Egan reports aimed to bring about a change in attitude from an adversarial and fragmented construction sector to one that is more trusting and better integrated. The publications express similar sentiments contained in earlier reports by Simon (1944), Phillips (1950) and Emmerson (1962), which also argued for better communication and more effective interaction between project participants. In many respects, the underlying message has not changed, but the context, technologies and the language have. The government reports have inspired many books, reports and articles that present a very positive argument for relational (interdisciplinary and collaborative) forms of working; examples being Baden Hellard's *Project Partnering: Principle and Practice* (1995) and *Trusting the Team* by Bennett and Jayes (1995). The message is that the AEC (architectural, engineering and construction) sector needs to move from 'segregated' teams to 'integrated' teams to improve performance.

Integration

The word integration is used to describe the intermixing of individuals, groups or teams who were previously segregated. Integration within temporary project organisations (TPOs) can occur on a number of different levels, from seeing the whole project process as an integrated one, to viewing the concept simply bringing together two separate work packages. The term 'integrated teams' has come into widespread use in the AEC sector, which although tautological, seems to be used to describe TPOs that are comparatively more integrated than might otherwise be the case. The majority of the literature promotes a highly positive view of integrated teams, while failing to acknowledge the inherent sociological and psychological challenges (see Emmitt, 2010).

Integrated design, supply and production processes are facilitated by cooperative interdisciplinary working arrangements. Integrated teams encompass the skills, knowledge and experiences of a wide range of specialists, often working together as a virtual team from different physical locations. Multidisciplinary teams may be formed for one project only, or formed to work on consecutive projects. Although there has been a move towards more collaborative working arrangements based on the philosophy of project partnering and strategic alliances, it is difficult to see evidence of real integration; instead there are pockets of collaborative work within and between projects.

Focusing on integrated processes is only part of the challenge. It is also necessary to look at the individuals involved with the project and look at how integrated their contribution is. How and when, for example, are the contractor and main sub-contractors involved in the early design phases? Are they an integral part of the design decision-making process or are they merely invited to attend meetings and asked for their opinion. For real integration to work, there needs to be social parity between actors, which means that professional arrogance, stereotypical views of professionals and issues of status have to be put to one side, or confronted. It also means that, in many cases, project teams need to be restructured and the project culture redefined through the early discussion of values, e.g. via value management exercises.

Integration of design and construction activities can achieve significant benefits for all project stakeholders. Improvements in the quality of the service provided and the quality of the completed project, reduced programme duration, reduced costs, improved value and improved profits are some of the benefits. Traditional procurement practices are known to perpetuate adversarial behaviour and tend to have a negative impact on the product development process and hence the project outcomes. Focus has tended to be on limiting exposure to risk and avoiding blame at the expense of creativity and innovation. The creation and maintenance of dynamic and integrated teams is a challenge in such a risk adverse environment. Fostering collaboration and learning within the project frame requires a more integrated approach in which all stakeholders accept responsibility for their collective actions. Project partnering is one approach that can help to bring the actors together, which combined with value management techniques, can help the project collaborators to deliver value to the client. Using new technologies and new approaches, such as off-site production, is another approach, which may change relationships (for example, the manufacturer may take the place of the main contractor). Similarly, the use of collaborative (relational) contracts and integrated project delivery approaches can help to provide value to client and fellow project participants.

Collaboration

Collaborative working takes place within organisations and within projects. Within organisations, work may be within a department or take place across departments. Collaborative working in projects is usually achieved through a coalition of disciplinary groups and teams and multi-disciplinary groups and teams. This dynamic and temporary organisation is usually referred to as a project team, or more accurately a temporary project organisation or a temporary project coalition. TPOs exist for the sole purpose of delivering a project within a specific set of parameters. They comprise a loose coalition of multi-skilled individuals with varying values, attitudes and goals. This temporary alliance is held together by legal contracts and the desire of the participants to achieve a shared objective. The quality of the project (the process) and the quality of the resultant building (the product) will be affected by the way in which these organisations and individuals interface and the effectiveness of the working relationships that develop over the course of the project. It may be an obvious statement to make, but the better the collaboration, the better the outcome of the project.

The word collaboration (or collaborative) is sometimes used as substitute for participation (or participative) and vice versa. Participation usually involves various levels of responsibility and power, thus some participants' interaction will carry more weight compared to others. Participation is the act of sharing or taking part in group decision making processes. In the context of a project, participative processes should include a wide range of stakeholders, from building sponsor and the professionals working on the project through to the building users and representatives of the local community in which the building is located. The intention of a participatory process is to achieve a higher degree of synergy by bringing multi-disciplinary actors together to share their knowledge, hopefully resulting in an outcome that would not be possible working individually. Outcomes are by consensus and the group members share ownership of 'their' decisions. Sometimes actors have equal participation rights, although it is more common for participants to have different levels of responsibility within the project.

Collaborative processes are concerned with equal participation, equitable power and shared decision making responsibilities. This is the philosophy of relational contracting such as partnering. Collaboration refers to cooperation with others, the uniting of labour to achieve a common objective. A collaborative project is one that involves multiple actors who work together and hence are mutually dependent upon one another. Collaborators should be prepared to listen to others, treat their ideas with respect and give each actor equal decision-making power. The aim is to resolve problems more effectively and also produce better outcomes compared to those likely via non-collaborative approaches. This often means that individuals and the organisations that they represent may have to relinquish power; which can be difficult for some participants to deal with, especially if they are more oriented towards a conflict based approach to business. It follows that it is useful to recognise and collaborative processes may not suit everyone, thus care is required when selecting the members of the temporary project organisation.

Collaborative design

Building design is rarely the product of one person's thinking process; rather it is the result of many different disciplines' collective knowledge (see also Chapter 2). Although it is possible to design a building by working independently, this is rarely the case in practice as professionals exchange information and ideas, discuss, negotiate and agree to a collective building design. However, for a true collective design process it is necessary for the project contributors to collaborate and develop the design together, a process that has been made somewhat easier by the development of information communication technologies and single virtual models (see Chapter 6 and 7).

Collaborative technologies

Information technologies continue to transform the way in which buildings are designed, manufactured, assembled and used. Improvements in the visualisation of designs, modelling, and communication between the participants has helped to provide a better understanding of the management of intricate processes. In particular, the development of IT and ICT such as project websites has made it easier to work concurrently and collaboratively from remote locations. 4D CAD models and building information modelling (BIM) provide the means for addressing the fourth dimension, time. These virtual models provide the interconnection between design information and the planning and scheduling activities (3D + time), providing animations of construction sequences. BIM provides users, regardless of their physical location, with the opportunity of testing, revising, rejecting and accepting design ideas in real time, i.e. it provides the means for collaborative design. BIM also provides a tool for improving the efficiency of communication within the TPO, since it is able to handle the vast amount of information required for coordination. Technical concerns over interoperability of various software packages and the availability of bandwidth to allow the large volume of data traffic to flow smoothly are ongoing concerns, but are being addressed. There is a perception (promoted by the vendors of the software) that better software will lead to better designs and better management of projects, although there appears to be a discrepancy between what the vendors claim and what happens in practice (Otter and Emmitt 2007). We need to remember, for the time being at least, that people, not software, manage construction projects.

Collaborative procurement

Although there has been an increase in the number of AEC projects that use relational forms of contracting (such as project partnering and integrated project delivery), it is still relatively modest compared to the more traditional approaches that rely on competitive tendering. Thus the majority of projects are still conducted in ways that, on the surface at least, are based on distrust and non-integrated working. This does not necessarily mean that the majority of projects are less efficient or less effective compared to those conducted in a spirit of trust and collaboration, but simply means that the project philosophy, i.e. attitudes of the participants, are different. The point to make is that the procurement route and the contract(s) used will help to colour the behaviour of the project participants. To promote collaborative working and collaborative design, management requires an appropriate contract.

The desire and ability to work in a collaborative manner varies depending on the procurement route and the legal contracts employed. This relates to risk avoidance and risk transfer. Collaborative procurement is often defined as a means of seeking value for money through a partnership of purchasers and suppliers. For example, local councils may work together to jointly purchase goods and services, and in doing so save money. In the AEC sector, the term collaborative procurement is also known as relational contracting or relation-based procurement, examples being partnering, joint ventures and (strategic) alliances. These are usually based on negotiated contracts that rely on open communication, transparent transactions and trust; the core values of collaborative working and also values essential to the adoption and effective application of BIM.

Case study 1A

The challenges of working collaboratively from a design manager's perspective

Paula Bleanch, Northumbria University

Before I took up an academic position at Northumbria University, I worked as a design manager, first for main contractors and then for an architect's practice. I often tell my students that this gave me a view of what it is like on 'both sides of the fence'. The role when working as a design manager for a contractor and a designer were similar, although the means were perhaps a little different. However, the aim was fundamentally the same and the way I approached my job did not alter between employers. What changed around me were my colleagues' attitudes. I found that when I worked for a contractor, my own team of builders, project managers and quantity surveyors would often accuse me of being 'too much on side with the designers'. On the other hand, when I 'changed sides' and worked for an architectural practice I was told that I cared too much about construction, and that working out how to build our design was not our job!

From my point of view I was doing the same thing, making decisions that put the project first. Sometimes this did not tally with my own personal interests, or indeed with my employer's interest (either designers or builders), however, I would say that working for the right outcome means supporting the project and this does not usually result in keeping all of the people happy all of the time. In fact, as a design manager I do not think that it is possible to make everyone happy, and nor should you aim to. Design is about commitment, compromise and trying to find 'win-win' solutions, but sometimes one has to have the courage to tell your colleagues, or fellow design team members, things they do not want to hear in order to get the right outcome for the building project.

For my undergraduate thesis, I investigated whether construction and design team members' traditional attitudes affected the design management process. You may not be surprised to learn that the design managers I interviewed did indeed find that attitudes affected their work, often in a negative way. We know that there is conflict in construction projects, but this is no surprise. But the really interesting question in the design versus build conflict is not about how it affects the design manager, although that's important. The important question is why the conflict exists at all. Why do we have a 'fence' in the first place? Why are there different 'sides' to change to? In my research I considered the barriers to effective design. And I should not have been surprised to find that the barriers to effective design management are the same as the barriers to effective anything in construction, namely, that we are a fragmented sector with many different people working on complex projects. In no other industry is the responsibility for design and construction so far away from each other. We are educated in silos (rarely do architects and contractors interact during their formative years in higher education) and hence we learn, indeed expect, to work within the closed borders of our disciplines. We expect conflict and we get it. Modern procurement routes, for the most part, do not seem to help erase the adversarial culture of the industry either. We waste a lot of time and energy resolving arguments. The basis of these arguments in my experience often comes from a perception in the project about what individuals 'expect' from the other disciplines. So rather than listening to one another, perhaps design team members sometimes hear what they want to hear, or what they expect to hear, rather than what is actually being said.

So how can we break down these barriers and become more collaborative in everything we do? Certainly BIM seems like an important part of the answer, which could lead to better integration of the team and perhaps even a blurring of the traditional roles, both in education and in industry. I do not think this just means less conflict due to a reduction in clashes between components and better coordination. I would like to see the building information model as a means of communication and collaboration for the team; a 'safe territory' where different ideas can be tried out before anything happens on site. I am not saying that BIM is a panacea for everything that is wrong with construction (anyone who says that is selling something) but I do think that it is a great start. I think it is certainly a good way of enabling change and that is what the government and clients are banking on too.

My hope is that 15 years into the future nobody will talk about fences, sides or silos anymore because they will not exist. Will it happen? For an answer to that question, it is best to ask the students who are graduating today because they will be the ones leading the change – and I hope they do a better job of collaborating than we did.

Collaborative communication

Communication is central to all activities and takes on increased importance when working collaboratively. Although much focus has been on the potential of information technologies to transform the way in which we work and communicate, there is still a need to communicate face-to-face and interact in meetings; which can be acted out in a physical space or at a distance using information communication technologies. Communication is dependent on the willingness of all participants to act and react, to listen and share as well as develop their skills for using communication effectively (Forsyth, 2006). It follows that project communication is likely to be most effective when all members contribute using the available communication media in the same way, and as agreed to at the start of the project, i.e. they follow project communication protocols. Experience tends to suggest that this is an ideal, rather than a reflection of reality. Both managers and project contributors need to understand and agree to systematic communications based on collectively agreed rules for a specific project context, and this should bring about both individual and collective benefits for the contributors (Otter and Emmitt, 2007).

Communication in any group has social and task dimensions. Task roles are those that determine the selection and definition of common goals, and the working toward solutions to those goals. Socio-emotional roles focus on the development and maintenance of personal relationships among group members. Open exchanges of information and sharing task responsibilities are essential for effective working relationships. Interaction that builds and maintains the fragile professional relationships within a project that are necessary to accomplish tasks is fundamental to project success.

Synchronous and asynchronous communication

Collaborative communication can be explained as a series of interactions between a group of participants using a web of communication channels and an assortment of media and tools. Communication involves some form of interaction between the sender and the receiver of the message via synchronous and/or asynchronous communication:

- Synchronous communication is when individuals and groups communicate at the same time through face-to-face dialogue and interaction in meetings, telephone conversations and video/online conferencing. Synchronous interaction is essential for addressing contentious issues, problem solving, and conflict resolution, for exploring values, developing trust and building relationships.
- Asynchronous communication is the term used when parties do not communicate at the same time, e.g. via email, sms (text messaging) and through intranets, by post and facsimile. Messages are sent and (hopefully) responded to sometime later. Asynchronous communication is essential for transferring a large amount of information to the receiver(s) when an instant response in not required or is not possible. Receivers of the message will be able to respond (if they feel they need to) at a later date once they have had time to assimilate the information and consider their response.

In a project environment it is common to use both synchronous and asynchronous communication, depending on the task, the preferences of the participants and the stage of the project. For example, face-to-face communication through dialogues and meetings may be used extensively in the early design phases when exploring possibilities and when consensus is necessary in order for the design to proceed. Later in the project, for example during the production information stage, the emphasis shifts to producing information and information exchange, so the emphasis is more on asynchronous communication. The challenge appears to be associated with identifying how design team members prefer to communicate within design and construction projects. According to Otter and Emmitt (2007) effective design teams use a balanced mix of synchronous and asynchronous communication.

Leading and stimulating effective communication is a challenging task for a number of reasons. First, the number of electronic tools is increasing and therefore both users and managers need to develop specific skills for their collective use (Otter, 2005). Second, there are differences between the participant's organisation's use of electronic information systems, combined with the variety of communication practices and this may create problems with compatibility. Third, there are differences in opinions and understanding on an individual level, including differences in the use of specific ICTs. Combined with the lack of a collective framework for meaning (Mulder, 2004), these factors can lead to misunderstanding within the TPO.

Communication breakdown can occur for a variety of reasons. Some of the most common errors relate to misunderstanding, the failure to communicate at an appropriate time (or failure to communicate at all), the wrong people communicating and the failure to ask questions (Emmitt and Gorse, 2007). Time pressures are a constant threat to effective communications; everyone appears to want answers and information immediately, and this can lead people into issuing information before it has been checked and/or saying something without first checking that the message is accurate. To a certain extent, good managerial frameworks and the appointment of individuals who understand one another will help to limit communication breakdown, but it will happen to lesser or greater extents regardless of how organized the TPO is.

Communication levels

One way of helping to identify communicators and communication routes is to analyse the TPO in terms of communication levels and communication channels (discussed later). Communication can be separated into five levels (Emmitt and Gorse, 2003), from the most private (intrapersonal) to the most public (mass communication).

- Intrapersonal (private and hidden)
  Intrapersonal personal communication is an internal communication process (cognition), which allows individuals to process information. Only one person is involved in this thought process, which is usually private and hidden from others. This is sometimes referred to the 'black box' because our thoughts and reasoning are not accessible to others, nor for that matter are our hidden personal and organisational agendas.
- Interpersonal (intimate)
  Interpersonal communication is between two people, a dialogue, which allows individuals to establish, maintain and develop relationships. Conversation is an intimate process during which common ground can be established quickly through the common understanding of terminology and language. It is also through interpersonal communication that we tend to make judgments regarding the trustworthiness of others.
- Group (familiar)
  Group communication occurs in small groups or teams, which may be disciplinary or multi-disciplinary in their constitution. These small groupings of individuals are usually able to develop effective communication quite quickly as they work toward a common objective. Groups are usually, but not exclusively, based in the same organisation and so they are familiar with

the organisational culture and protocols. An exception would be a creative cluster, a grouping of individuals from different disciplines and organisations assembled to engage in creative problem solving.

- Multi-group (unfamiliar)
  Multi-group communication occurs within social systems such as organisations and TPOs. With the exception of small organisations and TPOs, these social systems function because of the collective efforts of interdependent groups and teams, working to achieve a common goal, which is also termed 'inter-group' communication. The manner in which these groups and teams interact, and their familiarity with the other organisational cultures and protocols, will influence the effectiveness of their communications. In large organisations, projects may span departments and sub-groupings of staff. In TPOs, small groups and teams communicate with other, often unfamiliar, groupings, only getting to know how they like to work as the project develops.
- Mass (public)
  Mass communication involves sending a message to large audiences. This may involve advertising through the mass media (professional journals, newspapers, television, radio, Internet) or the dissemination of important information, for example changes to health and safety legislation, to the entire project team, via the project intranet.

All five levels serve different functions and all are equally important in achieving effective communication and in limiting the amount of ineffective communication. In the context of this book, the most pertinent communication level is multi-group communication or inter-organisational communication. In multidisciplinary teams, members come from different organisations, which have different organisational cultures and which also use a variety of information systems. Individuals also have different levels of understanding, opinions, skills and rates of adoption of communication technologies, as well as preferences for specific means of communication (e.g. Tuckman, 1977; Gorse, 2002).

Communication channels

Another means of analysing communication within the TPO is to look at formal and informal channels of communication:

- Formal communication channels are those set up and reinforced by the procurement route and legal contracts. Individuals are expected to follow certain protocols and use established communication channels that have been established for the project. These may be very similar, or subtly different, to the protocols and communication channels used within each contributing organisation. This means that individuals may have to adjust how they communicate across different projects. Formal communication routes can be designed prior to the commencement of the project and mapped during the project to ensure that they are appropriate for the project context and project stage.

- Informal communication routes are those that develop around the formal channels and are established by communicators to help them to do their job more effectively. Research has shown that in times of crisis project participants tend to favour informal communication routes to try and resolve the problem, conscious that formal communications are recorded and may be used as evidence if the problem turns into a dispute. The challenge for managers is that informal communication channels by their very nature are difficult to track and individuals tend to circumnavigate formal procedures, such as those set down as part of a quality assurance (QA) plan.

Communication networks

Communication channels form part of the communication networks that exist within organisations and within the TPO. It is common practice to use network analysis techniques to determine the communication structure within a social system, from which communication networks may be represented graphically. This is expressed by linking nodes (communicators) with other nodes, with the most frequent communicators represented by larger nodes and the less frequent by the smallest node. Analysis of the frequency of communication between individuals also allows the most active communication channels to be identified, usually represented as lines of varying thickness to represent frequency of communication. These techniques can be useful in helping to map communication routes and hence manage communications more efficiently. The disadvantage of network analysis is that it does not identify the quality of the communication, merely the quantity. Another challenge with network analysis is that the results tend to represent a network at a fixed point in time, the time the analysis was conducted, and do not address the change in the network over time (Rogers and Kinkaid, 1981). According to Rogers and Kinkaid, communication networks are so fleeting that they cannot be accurately mapped. This is especially true of AEC projects, where the social system is highly dynamic and fragile, thus network analysis would need to be conducted very frequently to capture the extent of the connections. To some extent, ICTs can provide a good picture of communication activity within a project environment, but the problem is that ICTs do not capture the informal communication that happens around the edge of the formal communication channels. Therefore our network maps can provide a reasonable indication of project communications, but cannot show the full extent of communication within the project network.

An evolving context for collaboration

Given the long timescales associated with AEC projects from inception to completion, it is necessary to recognise that the context is changing during the life of a project. The word 'context' comes from the Latin *contexere* (weaving together) and is used to describe the setting of an event or building: the weaving together of people and physical artefacts. Context is a crucial determinant of project success, as is the choice of appropriate processes, people, methods and tools. Understanding the context of a specific project is fundamental to the design process. Most obvious are the physical changes that happen on the site, since these are visible and can be quite rapid in the case of demolition and the erection of prefabricated elements but other, more subtle, changes occur as organisations and individuals enter and leave the TPO. Time also has a role to play, with local and global events (such as localised flooding, a major shift in the economy affecting interest rates and confidence levels, and global terrorist attacks) influencing the decisions made within projects. Client context might also change, especially on projects with a long duration since client organisations are dynamic, and it is quite probable that the main client contact person will change because he or she will be promoted within the organisation, or may move to another employer. Contributing organisations might also evolve, growing, shrinking or even going out of business, bringing about changes in the composition of the TPO as the individual contributors are replaced.

Social context

Any social situation is a sort of reality agreed upon by those participating in it, or more exactly those who define the situation. This is known as social constructionism, which relates to the way in which individuals make meaning from their social situation. Everyone who enters a situation does so with preconceived definitions of what is expected of him/her and the other participants. Such beliefs, which include expected interactions, are established from experience of previous groups to which the individuals have previously belonged. Thus, each situation confronts the participant with specific expectations and demands. Such circumstances generally work because most of the time our perceptions and expectations of important situations coincide approximately. Culture, society and individual power plays an important part in the norms that govern the way social groups act. The same aspects will also influence the behaviour of professional groups, but the group may also draw on those members with nominated roles, perceived expertise, skill and experience to determine who is allowed to interact, make decisions and play leading roles. However, some people, regardless of professional skill or experience, can use communication techniques to exert influence enabling them to gain power over elements of group behaviour and decision making.

Client context

AEC projects involve interplay between two complex and highly dynamic systems: the client organisation and the participants of the TPO. The term client is commonly used to describe the person(s) or the organisation(s) funding a project. Alternatively, terms such as building sponsor and customer may be used to describe the same thing. It is the client, be they an individual or a large corporation, who funds the project, usually by borrowing money. Therefore a number of other parties, such as banks, private equity investors, pension funds and insurance companies, will have a financial stake in the project. In some projects, these stakeholders may participate directly, with investors represented in strategic project meetings. Alternatively, the investors can indirectly influence the project by laying down a number of performance criteria to be met, for example stringent security measures, which will be incorporated into the strategic and project briefing documents. Understanding how each project is funded and mapping the financial stakeholders may be beneficial in helping to identify some of the project values, although clients are often reluctant to disclose such commercially sensitive information.

Clients may be defined in relation to their previous experience of construction (Emmitt, 2007a):

- First–time clients. The first-time client will need guiding through the design and construction process. This client may only commission design and construction services once, for example a house owner wishing to provide more space for a growing family. Here the brief will be a bespoke document. Effort may be required to ensure that the client fully understands the implications of the decisions being made. Visualisation techniques and simple graphics are very helpful in this regard.
- Occasional clients. The term occasional client tends to be used to describe people or organisations that commission design and construction services on a relatively infrequent basis. The gap between commissions may be lengthy and the type of project may be very different from the first. Thus learning from previous project experience may be challenging. The brief is likely to be a bespoke document.
- Repeat clients. Repeat clients tend to be major institutions, businesses and organisations with a large property portfolio. Typical repeat clients as examples would be food retail businesses and hotel chains. Here the commissioning of buildings is more likely to be part of a strategic procurement strategy, closely linked to the business objective of the organisation and their facility/asset management strategies. There may be an opportunity to establish integrated teams that move and learn from one project to the next. Similarly, the ability to make improvements to how design and construction activities are managed is also present with repeat clients. With repeat clients, the brief may include elements common to previous projects which represent the values and knowledge of the client organisation and codified in a standard brief.

Physicality and social interaction

In building design, a lot of emphasis is, not surprisingly, placed on the site, the genius of the place (*genius loci*). Every site is defined by its physical, three-dimensional context. Each will have its own particular ground conditions, its own micro-climate, its unique juxtaposition with neighbouring buildings, roads and boundaries. Sites will also have specific design constraints, such as access points, availability and proximity of services, existing levels, trees and hedgerows, etc. Sensitivity to the local environment, the pattern or grain of an area, combined with the physical characteristics of a specific site serve as design generators, with some taking greater priority than others in the design process.

Physicality of the site will set the stage for social interaction, helping to determine some of the project's stakeholders. This will include the legal owner(s) of the land who may, or may not, be the project sponsor, and their legal representatives. Selection of the temporary project organisation may also be influenced by the geographical location of the site, for example the appointment of an architect situated within the locale and the use of local suppliers and contractors. Thus local knowledge can be utilised while at the same time helping to stimulate the local economy and possibly reduce the environmental impact of the project by reducing, for example, transportation costs. Immediate neighbours will, quite rightly, have an interest in any development that might affect them, and they will express their views directly to the client or client's project manager, and indirectly through the town planning authorities. Similarly, local user groups and pressure groups are likely to take an interest, either in supporting or resisting the proposal(s) for the site.

Location of the site will also determine which local councils are 'responsible' for that site, i.e. whose jurisdiction the site falls under. Most obvious are the town planning department and the building control department, but other departments such as environmental health, refuse, highways, drainage, police and the local fire department will all become stakeholders in the project (if only for a short period of time), represented by one or more representatives of that department. Bylaws differ from town to town and the interpretation and application of legislation and guidance documents may well vary from one region to another.

The extent to which these disparate participants are included in the design process, and hence the project, will depend on the amount of inclusivity permitted, or encouraged, by the project team. Some clients and designers are happy to work with neighbours and the relevant authorities to help develop the design and acquire the necessary approvals as quickly as possible with as little conflict as possible. Others may be less keen to have their ideas compromised by officials and will adopt a less inclusive, often confrontational, stance. It is the attitude towards the various stakeholders that will determine the membership status of these individuals, as fully participating members of the project or merely occasional visitors to a small number of meetings. This assortment of individuals, groups, departments and organisations will all have a stake in the project, but it is important to recognise that they will not be party to any formal contracts relating to the project *per se*. Some of these may be reluctant stakeholders, involved simply because a neighbour has embarked on a new development which is perceived to be problematic.

Conclusion

It is only through a better appreciation of the benefits of collaborative working and the adoption of a positive attitude to collaborative ventures that the AEC sector will make real improvements in performance. As noted in the case study, the potential is there and this may be realised by the adoption of new technologies and new types of contract. However, the role of education in setting the mind set of those entering the construction sector, and also the role of organisations in promoting collaborative working within their organisations and within their projects cannot be underestimated. This demands clear direction and leadership by managers and it is an area in which design managers could, and should, be able to take a lead in establishing the right environment for collaborative design management.

Further reading

For further information on collaborative working see *Collaborative Working in Construction*, edited by Dino Bouchlaghem (Routledge, 2012). The fundamental issues underlying all projects can be found in *Managing Interdisciplinary Projects* by Stephen Emmitt (Routledge, 2010). Similar issues are addressed in Anthony Walker's *Organisational Behaviour in Construction* (Wiley-Blackwell, 2011).

2

DESIGN

Design has value to organisations as well as to building owners and users. Understanding the value of design to an organisation and its clients is fundamental to a successful business. It follows that some attempt to manage design as an activity, and design as an output, may be beneficial in helping to maximise design value for all stakeholders. Design is a highly creative activity, involving determination and compromise to realise the best value for a given challenge. Co-creation (or co-evolution) of design involves interaction between the project stakeholders, end users and society. The result of the design process is a collection of information – comprising drawings, schedules, specifications, calculations, and physical and virtual models – which is translated into a physical form by manufacturers, contractors and building operatives. It is crucial that everyone involved in managing building design understands what is being managed; and this starts with an understanding of 'design', a word that carries many meanings to many people.

Design as an activity

The verb 'design' is concerned with doing various activities we call designing. When designing, we are using cognitive and social skills, thinking, interacting, communicating and making decisions to resolve uncertainty and coordinate interdependent activities to avoid clashes or ambiguity. It is the interdependency of design work that leads to collaborative work, helping a wide variety of disciplines to harness expert design knowledge. Design activities, by their very nature are concerned with identifying problems, problem framing, proposing a variety of solutions and making choices. The manner in which designers address design problems has received much attention, (see for example Rowe, 1987; Brawne, 1992; Jones, 1992; Cross, 2011). According to Schön (1983) there are two schools of thought when it comes to design, either a rational problem-solving activity or an inter-active inquiry:

- *A rational activity.* The work on design methods was grounded in the view that design could be approached as a rational, systematic, problem-solving activity (and in some cases an information processing activity). This school of thought is still prevalent in many built environment programmes because in a simplistic manner it can help students to understand how design should be conducted and managed. Unfortunately, this over simplistic, one might argue 'technocratic', approach to design does not reflect the reality of how designers behave and makes very little attempt to recognise the 'messiness' of a highly creative and often ill-defined task. This misunderstanding is further reinforced by literal readings of process models, which imply a linear process.

- *An interactive inquiry.* The second school of thought is that design is a creative, artistic and interactive inquiry, strongly grounded in social, cultural and psychological thinking; what Schön refers to as the reflective practitioner. Architecture, and some of the architectural technology and architectural engineering programmes (and the authors of this book), subscribe to this view. From the outside looking in, to the uninitiated, the design process may appear to be rather chaotic and muddled; and hence difficult to comprehend and hence manage. However, for the designers, there is a clear understanding that process models (such as the RIBA Plan of Work) are there to guide the designer and are not to be followed to the letter.

These two schools of thought extend into education and the separation of the professions, with architects and the more creative architectural engineers, architectural technologists and engineers taking a very different approach to design compared to builders, contractors and the more technically orientated professionals. This often leads to misunderstandings about how long it takes to design and why certain aspects of a design are more important than others. And if these misunderstandings are not aired, it is highly likely that there will be requests for design changes somewhere along the project timeframe. Needless to say this can cause problems for design managers if they do not understand which approaches are being used, and by whom. This calls for design managers to be sensitive to the different approaches taken to design by different professional groups; and by implication design managers need to be experienced designers (see for example Case study 2A for a more detailed explanation). This is particularly important in a multi-disciplinary, collaborative, working environment in which the strengths of each discipline (and strength of different approaches to design) fuse in an attempt to find the best solution to a given design problem.

Types of design problems

Rowe (1987) makes a useful distinction between the different types of design problems that may confront a designer:

- Well-defined problems. Where the goals are clearly prescribed, an example could be a repeat building type, such as a drive-through fast-food unit, where the building design is already known. The unknowns may relate to site conditions and the uncertainties of the town planning process. Other examples of well-defined problems are standard retail units (e.g. supermarkets), speculative office buildings and 'standard' house designs as employed by the speculative housing developers.

- Ill-defined problems. These are common to most architectural and urban design problems where the means and the goals are not clear at the outset. The client knows that they want to commission a building project, but there is little clarity about exactly what is required at the outset. Here the designer's role is to establish the nature of the problem (problem definition and subsequent redefinition) and clarify the goals of the client and building users. This is usually done through briefing and sometimes by starting to design to explore possibilities and preferences, and hence reduce uncertainties. For example, a client may want a new house built on a plot of land, but not have too many clear requirements at the outset other than the type and number of rooms required. This is where the skill of the designer comes into play in helping to tease out what the client really wants and then deliver value through a creative and functional design solution.
- Wicked problems. When a design problem is extremely difficult to define it is referred to as a wicked problem (Churchman, 1967). In this situation, the problem defies easy formulation and definition, requiring constant reformulation and definition. Here it is common to have a very vague brief and to start designing to try and establish the nature of the problem. And given that the problem also lacks clear goals, there is no definitive stopping point, (solution); therefore designs may have to be continually reworked and it is not always obvious if solutions are appropriate or not. In such situations, it would be possible to carry on designing for an indefinite period, but usually the constraints of time and finance force a decision to be made.

One of the challenges for design managers is to first establish the type of problem they are dealing with and then apply the appropriate managerial frames to suit the context. This is usually done in the early interactions with clients and may form part of the briefing process; and in the case of wicked problems form part of the design process. Managerial techniques such as value management can also assist in helping to identify function and value (see Case Study 5A). Opportunities for designers to interact – collaborate, communicate, discuss, negotiate, make decisions and learn – must be included within the managerial framework.

Design as learning

Designers will readily attest to the fact that designing is rarely a straightforward activity, involving many changes, jumps and iterations in the process as we seek to move from the unknown (the problem) to the known (the solution). This often muddled process will result in collective learning, with new knowledge generated for individuals and for their employers. One of the challenges for organisations is to capture that knowledge for future use, encoded in office standards and procedures, and captured via, for example, podcasts (see Case study 6A). The challenge for designers is not to 're-invent the wheel' every time they design, usually because time is at a premium and shortcuts have to be taken to stay on programme. Familiar processes are followed to help tackle the problem, with individuals developing their own ways of working, referring to their favourite sources of information, and relying on their favourite solutions to familiar design problems (Wade, 1977; Mackinder, and Marvin, 1982; Emmitt, 1997). The problem with relying on familiar solutions is that we fail to learn and we fail to innovate; and in the worst cases we cease to be designers. Therefore it is essential that design managers allocate appropriate time to the design process to allow the designers the space to explore new ways of solving the problem before them.

Creative disorder, chaos, messiness, jumping in at the deep end, are essential and familiar characteristics of designing (Jones, 1992). Models and frameworks used for managing designers must allow some freedom to explore and experiment, otherwise creativity will be stifled and designers will be frustrated. Frameworks and models must also provide opportunities for designers to reflect and learn from the process in order that they are better equipped to tackle new design problems. This can be done within the office as part of the organisation's commitment to continual learning and also within the remit of the project. This may be done as the project proceeds (through-project learning) and/or very soon after the completion of the project (post-project learning). If learning events are not built into the design management process, the project participants will fail to learn from their collective experiences (see Emmitt 2007a).

Design as an output

The noun 'design' describes the result of designing, which is represented by an artefact, be it information or a physical (or virtual) product. Design results in a product (or process) that did not previously exist, for example a new building or bridge. In the construction sector, the output of the co-creation of design is a set of information from which the contractor builds the design. This (design) information comprises a number of artefacts, primarily drawings, specifications, schedules and models, in which design intent is codified. Information is central to both the design and construction process. Drawings and written documents are used to describe and define a construction project and designers need to understand the relationship between the written specification (which defines quality) and the drawings, i.e. what goes where. The challenge for those charged with realising the building is to translate this information into a physical artefact, safely and efficiently.

In some respects, the output of the design process (information, specifications, schedules and models) is easier to manage compared to the act of designing because managers are dealing with artefacts rather than people. Here the design manager's attention should be focused on the quality of the information provided by the designers, with particular attention paid to accuracy and the coordination of design information. Clash detection is an associated function, although one that is being addressed via computer software and the uptake of BIM.

At its best, this project information will be clear and concise and easily understood; so that the building will be completed on time and within the budget. At its worst, poorly conceived and shoddy design information will lead to confusion, inefficiency, delay, revised work, additional expense, disputes and claims.

Coordinating the output

The major bodies in the British construction sector established the Coordinating Committee for Project Information (CCPI) in an attempt to improve the quality of project information, and hence reduce uncertainty and disruption to construction activities. The CCPI has been instrumental in producing guidance for the design team, developing CAWS and contributing to SMM7. Central to the ethos of the CCPI is compatibility between the drawings, the written specifications and the bills of quantities, i.e. each artefact should complement the others when read as a set of information.

Coordinated project information (CPI) is a system that categorises drawings and written information (specifications) and is used in British Standards and in the measurement of building works, the Standard Method of Measurement (SMM7). This relates directly to the classification system used in the National Building Specification (NBS). One of the conventions of coordinated project information is the Common Arrangement of Work Sections for Building Works (CAWS), which has superseded the traditional subdivision of work by trade sections. CAWS list around 300 different classes of work according to the operatives who will do the work. This allows bills of quantities to be arranged according to CAWS.

Design quality

The artefact will be perceived and judged by a wide variety of people, ranging from critics to owners, users and passer-bys. Design quality is a subjective value, somewhat coloured by one's position, and therefore not easy to measure objectively. Despite that, a number of attempts have been made over the years to apply quantitative measures to design evaluation. Value engineering has a number of techniques that may be used to evaluate and prioritise design (such as the value tree and functional analysis). Another example is the design quality indicator (DQI), which provides the means to evaluate the design and construction of buildings (see www.dqi.org.uk). Although architects may find such approaches counter-intuitive, many clients and building professionals find it useful to put numbers against a number of options to help with the decision-making process and hence help give some meaning to an otherwise intangible issue.

The designers

Although design tasks are carried out by individuals, it is very rare for one individual to design a building. Buildings require the combined efforts of many individuals, working and designing collaboratively to provide value to their clients. In the construction sector, the primary building designers are the architects, architectural engineers, architectural technologists, landscape architects, interior designers, structural engineers and services engineers (covering a wide range of areas such as lighting, acoustics, water, heating and ventilation, fire and security). This diverse collection of designers also includes building product designers and manufacturers, specialist sub-contractors, and specialist contractors. Many other professionals and trades also contribute to the design of our built environment such as environmental consultants and indirectly the cost consultants, insurers and funders, value managers, facility managers and project managers. All of these individuals have personal values and operate within their organisational values; resulting in a wide ranging mix of views, understanding and approaches to design. The challenge with every project is to bring these stakeholders together in the most appropriate manner for a given project context so that the interfaces can be managed effectively and efficiently (see Emmitt, 2010). Bringing people together early in the project also helps to establish responsibilities, agree individual contributions and establish operating methods and means of communication before the project commences.

The challenge for managers is identifying the interfaces between these activities, which may be clear cut in some traditional forms of procurement and less so with collaborative approaches, and apply an appropriate managerial framework to help guide designers to creative and effective solutions. An associated issue relates to responsibility for design, which although covered by contracts, remains a contentious issue for many and can result in rather defensive behaviour in an attempt to avoid design liability. Openly discussing roles and responsibilities at the start of projects can assist in allocating design risk fairly within the temporary project organisation.

Case study 2A

Design management: the designer's perspective

Tanya Ross, Head of Design Project Management Group, Buro Happold

The process of design is creative, informal, dynamic…The best designers, those with real flair, are often perceived as woefully disorganised; their inventiveness an apparent product of haphazard thinking and random brilliance. In reality, design in the built environment is a cyclical process, with a number of different contributors from different disciplines collaborating to refine a design solution over a period of time. From the initial idea ('yes, we need a new building') the team will gradually hone the design to meet many (and varied) client and statutory requirements; while constantly testing their proposals against their own standards and vision for the project. Ultimately, the final solution should be the optimum synthesis that meets client aspirations, satisfies regulatory standards, and integrates the practical engineering needs with the architectural and aesthetic vision.

Managing such a process within a linear timeline is always going to be a challenge. At Buro Happold, we have a small team of Design Project Managers (DPMs) that support our large, complex, multi-disciplinary projects by managing design teams. Our DPM team does not employ recent graduates, since we believe that individuals need to really understand buildings and the construction industry before trying to manage designers. Our policy is to recruit experienced architects, engineers, and quantity surveyors as design managers – individuals with appropriate knowledge of design and the challenges associated with producing good design.

Design management is the application of a wide range of professional skills to coordinate and optimise the design process, and ensure that the delivery of a project meets or exceeds the expectations of both the client and the team as a whole. It is an essential function on all projects and of particular significance on complex, multidisciplinary projects, where clients are looking for effective and high quality delivery of a fully integrated design service. At Buro Happold, design management is applied from project initiation right through to project completion. Our work includes:

- advising on scope, procurement strategy, objectives and organisational structure;
- preparing design briefs and appointment details;
- establishing project financial structure and controls;
- developing and monitoring project execution plans and programmes;
- establishing communication routes and information management procedures;
- maintaining design programme reporting on progress and any corrective measures required;
- monitoring design development, advising on contractual, financial and functional implications of design;
- managing design discipline interfaces and design changes;
- coordinating design team output to the client, the construction team, approval authorities and other external parties.

Pulling together different disciplines successfully is more than a management process: it requires solution driven individuals with leadership and excellent communication skills. Understanding design interfaces and maintaining open communication between all parties is fundamental in harnessing the best from creative collaborations. The greatest benefits accrue when all team members are able to excel within their own discipline. Extra value may be found in: the speed of delivery, minimum conflict, reduced time on site, price certainty, as well as exceptional design flair.

The Millennium Dome

By way of illustrating the importance of design management I have chosen the Millennium Dome project in London, where the design manager became a pivotal role for the team. As a clear first point of contact, the design manager became an information hub, distributing incoming material to all the relevant parties, and ensuring that outgoing information was generated on time and to an appropriate level of detail. Tasks of the design manager included:

- developing early design programmes and construction programmes;
- agreeing tender documents for early packages (before contractor appointment);
- managing the tender process for early packages;
- interfacing with the landowner, English Partnerships, and their designated remediation engineers;
- interfacing with the Highways Agency (the Blackwall Tunnel runs under the Millennium Dome);
- liaison with the Environment Agency;
- point of contact for the local authority, including technical submissions;
- set up and chair weekly internal coordination meetings;
- represent the engineering team at client progress meetings;
- develop and agree design programmes to support the construction manager's construction programme (allowing sufficient time for the designers to complete the creative cycle);
- establish document control systems;
- maintain quality control systems;
- set up drawing review system;
- demonstrating intellectual property rights to the client;
- representing the project in the media – from local radio through to national TV.

Many of these tasks required a level of initiative and self motivation, and a not inconsiderable degree of patience.

The role of design manager is one that requires balance:

- As part of the engineering design team, to facilitate the designers' role, by establishing realistic target dates for deliverables and defending the engineers from unnecessary distractions to allow them to meet those target dates.
- As client liaison, to ensure that changes have been considered before they are passed on to the designers, with all the implications of a particular course of action recognised before a change is instructed.
- As point of contact with the statutory authorities to provide whatever information the various agencies and interested parties have demanded.
- As Buro Happold's project manager, to keep in touch with sub-consultants, ensuring that they were fully informed about their aspect of the works.
- As someone with an overview, making oneself available to the architect, client and construction manager, solving as many ad hoc problems as possible.
- And, of course, managing the resource planning, invoicing and fee control side of the team too.

Perhaps the most rewarding part of my role at the Millennium Dome was in helping some of the younger engineers deal with the stresses and strains of working long hours on a demanding site. Being asked to act as a sounding-board for ideas, or an initial management contact for a perceived problem, was tremendously satisfying, especially where I felt I was able to help find solutions to personal concerns. And one of the most exciting roles that I was invited to play was as an engineering representative to the media. With so much public interest in the Millennium Dome, I had the opportunity to speak on both national and local radio, as well as make appearances on national and local television. The possibility of introducing the engineer as a significant contributor to the built environment to a wider audience was very exciting, and although it required early mornings and late nights, I wouldn't have missed it for the world.

Once the fuss over the Millennium Dome was over, I took on some different challenges, leading a project to redevelop a degraded brownfield site in the north-west, and then acting as a sustainability consultant for the team master planning the Lower Lea valley in East London. However, when this work led to the successful bid for the London Olympics, I knew that what I really wanted to do was to lead the engineering team for the stadium. So I have been project director from the bid, from back in September 2006, right through to completion of the stadium base build in March 2011. The Olympics is a tremendous opportunity for the UK construction industry, and I think we have showcased what can be achieved with a collaborative approach and the right teams on board. The stadium, whatever may be written about it architecturally, meets its brief with minimalist elegance. This is not a show off stadium, but a carefully considered response to a challenging set of requirements which uses the minimum materials possible and it was completed early and under budget.

Currently, I lead the Design Project Management team at Buro Happold. We are nearly 20 people across the practice, both in the UK and overseas, all working to achieve harmony between the designers on large, multi-disciplinary projects. We aim to give our designers more space and time to design, by taking on some of the necessary but less creative aspects of their role. We also aim to give our clients better value, by ensuring efficient communication between the team and a clear point of contact for third parties. We must be doing something right, because our clients and our colleagues are happy.

Future directions

Since the construction of the Millennium Dome (1996–1999), things have moved on, not least in terms of the assistance provided by technology. The prevalence of email has speeded communications across even geographically dispersed teams, while the use of collaboration sites (A-site, Buzzsaw, 4Projects) and the emergence of BIM has greatly eased the sharing of information around the team. Mind you, these tools bring their own difficulties – as the issues over ownership of BIM models continue to occupy construction lawyers, while project extranets can be a mixed blessing. They are excellent as repositories of validated current project information, as rigorous version control ensures that only the most up-to-date drawing is available. They do require designers to be equally rigorous in maintaining their drawing lists, so that superseded drawings are withdrawn, and they can be clunky to navigate and frustratingly slow to upload or download. And of course they do not really work for the exchange of developing designs, the shaky pencil sketches of architectural 'what-ifs' that are the bread and butter of the design process. Another difficulty seems to be in getting the whole team to sign up to the designated system – individual companies QA systems or technological constraints can prove to be difficult obstacles to overcome, even when a client deems it a mandatory requirement. I do think that with improving access (bandwidth availability and more technologically savvy users) cloud-hosted project extranets and BIM will become the norm for the industry, and some industry-wide standards for their use and adoption must be agreed soon.

And our complex projects seem to have become ever more complex – increased legislation aimed at better quality buildings means that there are more compliance issues, and the emergence of increasing numbers of specialists (access consultants, IT consultants, sustainability consultants, BREEAM consultants, colour consultants, branding consultants, etc.) has led to even bigger teams. Many of our clients are now demanding spectacularly fast-track design and build programmes, in spite of the increased complexity, so the role of the design manager has increased in importance. An invaluable addition to a multi-disciplinary team is an experienced construction professional who can sit at the heart of the creative design team, giving them the freedom to design, while maintaining sufficient structure to retain control of the overall process.

Process guides

The process of designing buildings is typically conducted in professional offices, with collaboration taking place via individual projects; facilitated by digital technologies and physical interaction in meetings and workshops. Management of the designers will take place within individual offices, with overall project coordination overseen by a project manager(s) and overall project design overseen by the (construction) design manager(s). Process models and frameworks are essential to allow all designers to understand their position, roles and responsibilities in relation to others; they are also essential for design mangers.

A number of well known frameworks and models exist for the administration and management of projects, their suitability dependent on ensuring the best fit between the client, the aims of the project and the project participants' organisational needs. In the UK, the best known is the RIBA Plan of Work, developed by the Royal Institute of British Architects for use by architects. Other process models have been developed by others, such as the British Property Association and the generic process management protocol. With the uptake of BIM there will, no doubt, be the need for new process models that better reflect our digital working environments. Whatever process management tool is used it is crucial that all those working on the project are comfortable with it and understand the boundaries of their decision-making responsibilities; it also needs to be appropriate for the context of a given project. Early discussion about the process management tools being used can help to eliminate misunderstanding later in the project.

Models represent a rational (and often prescriptive) approach, although the reality is that considerable flexibility is required in practice. Frameworks help to give a degree of formality to sub-sets of work and help to formalise the interface of work and workers. The formality of the framework is such that it is sufficiently understood by those contributing to the process to enable informal interactions within and around the frame, i.e. it is interpreted liberally. It can be misleading to put too much emphasis on the framework model. It is more important that everyone understands the project, roles and responsibilities, which needs to be discussed at the outset of projects and re-visited at regular intervals during the course of the project.

Managerial frameworks should facilitate design work, communication, knowledge sharing and information flow. The process needs to be mapped and a suitable operating structure devised to manage the various activities in an efficient manner. It is a sensible policy to assemble the main actors, discuss how the project should be planned, and agree on the most appropriate framework in which to work collaboratively (a bottom-up approach). This tends to be a more effective approach than implementing a system and expecting everyone to be comfortable with it (a top-down approach). From the perspective of the architects, the model must allow space for spontaneity and creativity. Frameworks should be customised to suit the project context and as a minimum should contain:

- clearly defined stages, roles, tasks and responsibilities;
- value and risk management workshops at strategic intervals;
- project milestones;
- last responsible moment for decision making;
- control gateways to coincide with the end/start of different phases;
- learning opportunities and feedback loops.

RIBA Plan of Work

One of the best-known guides for managing design projects is the RIBA Plan of Work. This was first developed in the early 1960s by the Royal Institute of British Architects (RIBA) as a tool to help architects manage their projects, and has since been updated to reflect changes in how we build and more recently our attitude to our environment. The guide continues to be used extensively by architectural firms and has parallels with guides produced by architectural bodies in other countries. The Plan of Work has been criticised (mainly by non-architects) as being a linear model that promotes a segmented approach to the management of projects. Although this may appear to be the case when read literally, it fails to recognise the way in which architects use the model. The Plan of Work provides a familiar guide that helps designers navigate their way through highly complex and interwoven activities. The guide is not followed strictly, since design activity requires constant iteration and reflection as the design work develops. A fluid approach is possible because of the guide; forming a backbone to decision making and delivery of work to defined milestones.

The RIBA Plan of Work and similar frameworks give a formality to the process that is understood by other actors. The process is a number of sub-sets of formality (stages or phases) held together by a number of loose joints. It is the positioning and control of the joints that determine the framework of the work plan and the subsequent effectiveness of the process. The actors interpret formality differently, but there is enough common understanding to make the process work. Care is required when implementing new (unfamiliar) frameworks, as time is required for the actors to engage in communication and hence develop a common understanding. The main phases are the inception phase, the design phase, the detailing phase, the implementation phase and finally the post-occupancy phase.

Other guides

Other models are also available that claim to be less linear, although these, too, break down functions into discrete work packages and/or areas of responsibility. Process models can help to illustrate, or model, a web of relatively complex activities under a generic framework applicable to all projects. These models are more suited to large and complex projects and application to small projects may be unnecessary and inappropriate. Emphasis tends to be on integration of activities, concurrent development of work packages, knowledge transfer and change management. Value management and value-based management models are based on the discussion and agreement of values via facilitated workshops. Consensus and the creation of trust is a fundamental component of these models. Workshops start with team assembly and continue to project completion and feedback. Workshops encourage open communication and knowledge sharing while trying to respect and manage the chaotic nature of the design process. Cooperation, communication, knowledge sharing and learning as a group help to contribute to the clarification and confirmation of project values. Getting to know fellow actors and the development of trusting relationships is an essential feature of the model. The model is suited to partnering type arrangements and relies heavily on the skills of the process facilitator to drive the work forward.

Often the frameworks are implemented in a pure form, although it is also relatively common for architectural offices to take different elements from models to suit their work ethic. Flexibility and some degree of latitude for change is an essential requirement of good frameworks. Allowing some tolerance between well defined stages or works packages is a familiar and effective way of allowing for some degree of uncertainty. For example buffer management techniques have proven to be successful in helping to manage the interface between different work packages.

Common phases

Reviewing the various design process models reveals a number of similarities. These can be summarised as:

- *Inception and client briefing*
  Contractors and professional consultants will be vying for the attention of the organisation or person(s) funding the project. This is important for effective briefing and also for helping to secure repeat business from clients, for which there is considerable competition. As contractors have embraced design management, they have also become better equipped to interact with the client during the briefing stage; either conducting the briefing stage themselves or outsourcing it to designers or professional brief takers. It is through the briefing process that the users' wants and needs are explored and understanding is achieved. Failure to conduct a comprehensive briefing process will usually lead to design changes and cost uncertainty later in the project.
- *Design*
  The outcome of the briefing process, the brief, sets out what is required and informs the designers. At this stage, more people join the temporary project organisation, comprising managers, cost consultants and designers to explore solutions to the given challenge. During the development of the design, it is inevitable that it will change as designers and engineers work to maximise design value and minimise waste. Therefore some attempt must be made to deal with design changes as the design evolves. As a general rule of thumb, the later the design change occurs in the process, the greater the implication (cost and time) of accommodating that change. For schemes with a high degree of off-site manufacturing it is essential that the design is reviewed and approved prior to manufacturing commencing, although one could also extend this argument to more traditional forms of construction. In terms of efficiency and reducing uncertainty, it is necessary to complete the design information, review it and approve it before entering the construction phase. This allows for accurate bidding by contractors and sub-contractors and accurate construction.
- *Construction*
  The realisation of the abstract design into a physical form is achieved by interpreting the design ideals codified in drawings, calculations, text and models and converting that into an artefact. It is here that there is a change in culture and responsibilities. The construction design manager will be working alongside the project managers and the construction managers to ensure that design quality is maintained throughout the often long build period.

- *Maintenance and recovery*
  Buildings will require regular maintenance, repair and upgrading to suit changing user needs. Feedback from building maintenance managers and facility managers can be used for the development of new projects, helping to integrate important experience and knowledge of how buildings are used and how they perform over time. Once the building becomes obsolete, it will need to be remodelled or demolished and a new artefact put in its place: and the process starts once again, albeit from a different point in time.

The manner in which designers and design managers choose to tackle these phases will be determined to a certain extent by the characteristics of the project and the technologies available, but also by the characteristics of the designers and their preferred ways of working. Good design managers will be sensitive to these issues.

Conclusion

Once we understand design, we are better able to manage the design process and the designers collaborating to co-create stimulating and functional designs. Failure to appreciate the various skills and competencies that interface within a project environment will inevitably lead to poor design management and ineffective design solutions. On the other hand, appreciation of the different knowledge sets, skills and competencies that may contribute to a construction project will go some way in helping to facilitate greater understanding and better performance in both process and product. The case study provides a number of clues as to the value of good design management, and now the emphasis turns to the management of design and design management in Chapter 3.

3

DESIGN MANAGEMENT

Design management is the convergence of two cultures, the culture of design and management. Taken at face value, design and management may appear to be quite separate worlds, although there are many similarities between them if one looks beneath the surface. Design is concerned with shaping solutions to problems, as is management; the two should complement one another even though the language, terms and approaches used in each knowledge domain differ. Unless someone takes responsibility for ensuring design quality is delivered to the client, it is highly probable that there will be decisions made during the course of the project that (adversely?) affect design quality and hence the performance of the building in use. In this chapter, emphasis is on design management, which sets the scene for discussing the role of the (construction) design manager in Chapter 4.

Fields within fields

The term 'design management' is often used quite loosely to cover a wide range of functions, which vary between industrial sectors and their bodies of literature. According to Borja de Mozota (2003), design management originated in the UK in the 1960s and was initially used to refer to the management of the relationship between a design agency and its clients, with Farr introducing the term design manager in 1966. Arguments for managing design effectively can certainly be found in publications ranging back to the 1960s, for example in architecture (RIBA, 1962; Brunton et al., 1964), product and industrial design (e.g. Farr, 1966; Archer, 1967); with the construction literature picking up on the theme sometime later (Gray et al., 1994; Austin et al., 2001). Reviewing the literature, it would appear that the design management role has developed independently between the fields of product design, architecture, and construction; essentially fields within fields. It is not entirely clear why this has happened, although it is possible to speculate on some of the reasons.

An obvious factor is the different business context in which design managers operate in their well-defined industrial and professional sectors. What works in the field of fashion may not necessarily translate to construction: although there might be similarities to architecture. Another factor relates to the scarcity of literature, which makes it difficult for researchers and practitioners to know what design managers in other sectors are doing, and hence makes it difficult to learn from others.

Despite the separate evolution of the literature, it is possible to see parallels between the fields of knowledge, and most of the main concepts appear to permeate all three fields. Consistent within the literature is a shared understanding that design management concerns the business of design, and covers strategic and operational issues. The overriding message is that design can add value to a business if it is managed effectively. Obviously much of the literature draws on general project management concepts and techniques, although to see design management as merely a sub-set of project management would be a mistake.

In the concise review offered here, the origins of design management are explored from three perspectives; that of design management in product design (generic design management); the field of architectural management; and the construction design management discipline. The intention is not to provide narrow definitions, but to look for trends and overlaps between these fields of knowledge.

Generic design management

It is the field of industrial product design that design management is most established, with associated literature relating to brand management, fashion design and service design management. An early publication *Design Management* by Michael Farr (1966) put a convincing case for the design management role as an essential aid to attaining and maintaining competitive advantage. Farr (1966) made the observation that the design management role was not particularly well understood at the time of writing his book. His seminal book provided a comprehensive overview of design management and highlighted the need to manage design from a business perspective. The creation of the Design Management Institute (see www.dmi.org) in the United States in 1975 was instrumental in raising the profile of design and legitimising the role of the design manger and has helped to stimulate a growing knowledge base. Although the aim of the Design Management Institute (DMI) is to primarily serve senior design executives, it does play a role as a source of information for teachers and researchers (see Borja de Mozota, 2003).

In addition to the early work of Farr (1966) a number of authors have published books on design management, some of the best known being Oakley (1984), Cooper and Press (1995), Boyle (2003), Borja de Mozota (2003), and Best (2006, 2010). Although this body of work does not address design management from the perspective of an architect or construction professional, it is possible to see parallels with the management of design in AEC. Many of the generic approaches may, with care, be applied to the management of construction projects. The message within the generic design management literature is that design management is integral to new product development and business success; relates to the entire life cycle of the product and is a strategic asset.

An all embracing definition that has relevance to the construction sector can be found in Boyle's book *Design Project Management*: 'Design management involves understanding, coordinating and synthesising a wide range of inputs while working alongside a diverse cross-section of multidisciplinary colleagues' (Boyle, 2003). Of course, this observation could also be made of project management.

Architectural management

The term 'architectural management' first appeared in the 1960s in *Management Applied to Architectural Practice* (Brunton *et al.*, 1964). This book was a direct response to the RIBA's report *The Architect and His Office* in which architects were criticised for not adequately managing their businesses, and by implication not managing design. Brunton *et al.* were instrumental in making the link between the effective management of design at a project level and the effective management of design within the organisation; a theme echoed in Farr's seminal work on design management published two years later. However, it was not until the 1990s that the term started to be used more widely in the architectural literature (e.g. Nicholson, 1992, 1995; Emmitt, 1999a, 1999b), and an international research network dedicated to architectural management was established in the 1990s (see Nicholson, 1992; Emmitt *et al.*, 2009).

One of the curious features of the architectural literature is that the majority of publications address either the management of the architect's business (often referred to as office or practice management, see for example Sharp, 1986; Littlefield, 2005) *or* the management of individual projects (often referred to as project or job management, see for example Allinson, 1993, 1997; Green, 1995; Dalziel and Ostime, 2008). The first body of literature is concerned with the peculiarities of managing the professional service firm, the environment in which designers work. The second body of literature is concerned with managing projects, the vehicle by which designs are created and realised. Rarely do the authors of these two sub-sets of literature recognise the relationships between the office and the project.

Architectural management does address the interdependencies between the management of the professional office and the management of individual projects. It is the synergy between the business and the project portfolio that makes architectural management a unique subject (Brunton *et al.*, 1996; Emmitt,1999a, b, 2007a). It is within this very small body of work that the similarities between the generic design management literature and architectural management are to be found – in particular, the realisation that design management is a strategic asset.

It is not common for design managers to be found within architectural offices, but that does not mean that design is not managed; it is, but in a different way from contracting organisations (see Emmitt, 2007a).

Construction design management

As discussed in the Introduction, the term design manager (and building design manager) started to be used in the 1990s as procurement shifted in favour of design and build. Early examples of the literature included a seminal report by

Gray *et al.* (1994) and the resultant book *Building Design Management* (Gray and Hughes, 2001), in which the growing importance of the design manager in construction was emphasised. However, the need to improve design management techniques (Bibby, 2003) and to understand better the design management role from a contractor's perspective (Tzortzopoulos and Cooper, 2007) – and also in terms of sustainable construction (Mills and Glass, 2009; Rekola *et al.*, 2012) – illustrate the fact that the discipline is still developing.

Bibby (2003) noted that while there was growing interest in design management within the UK construction sector, there were a number of barriers to be addressed before design management could be successful. These barriers related to the nature of the design process and the construction practices at the time. Tzortzopoulos and Cooper (2007) found that there was a lack of clarity and understanding of the design manager role within the construction sector, supporting similar findings by Bibby (2003). However, as the role has matured and expanded, there has been increasing level of clarity in what the role covers (see also Chapter 4). Within the construction design management literature it appears that:

- Design management is implemented at the level of the project; it is not yet integral to the successful management of individual projects *and* the business (as advocated in the generic literature and the architectural management literature). However, the link between project and business effectiveness is inferred in a growing number of job advertisements for pre-construction and construction design managers.
- Design management is largely a coordinating and integration function, and does not yet relate to the entire life cycle of the product (built asset). However, this seems set to change as the role expands and BIM is adopted more widely.
- Design management is not yet considered by contractors to be a strategic business asset, although this is evolving rapidly and is likely to change as the understanding and application of design management matures.

Similar to the architectural management literature, this body of knowledge is small and still lacks clear terms and definitions. Although there is some variation in interpretation within the literature, it is evident that the construction design manager is responsible for managing design at a project level.

Leaner design management

Two management approaches that have been widely promoted in recent years are project partnering and lean construction. Both are concerned with process improvements and in their own way aim to improve integration between inter-disciplinary contributors through collaborative working. The ethos of project partnering is based on cooperative project relationships and long-term relationships in the case of strategic partnering (and project alliancing). The lean philosophy is one based on trust and shared responsibility for risk and reward – a philosophy central to collaborative working and closely related to integrated project delivery (IPD).

Lean thinking

Lean thinking has been taken from the manufacturing sector and adapted to the AEC sector, mainly under the guise of 'lean construction' (see Jørgensen and Emmitt, 2008). The ethos is to improve construction through lean processes and integrated project delivery, for which collaborative working is a fundamental characteristic. The philosophy is to reduce (process) waste and maximise value for the end customer via continual improvement. Adopting a lean approach also requires cooperative partnerships and a change in attitude to how projects are procured and delivered, facilitating greater interaction between design and construction (Jørgensen and Emmitt, 2009).

Originally developed in manufacturing, with Toyota being a successful exponent of the approach (Womack *et al.*, 1991; Womack and Jones, 1996), lean thinking has been applied to construction (for further information visit the Lean Construction Institute's homepage – www.leanconstruction.org). Most of the published research has concentrated on the construction phase, rather than the proceeding design activities, and literature relating to lean in the design management phase is relatively sparse (see for example Jørgensen and Emmitt, 2008; Tribelsky and Sacks, 2011). This is one of the curious anomalies of the lean construction field, since it is in the design phase that most value is generated and the majority of waste eliminated.

With close links to supply chain management, logistics, and total quality management, lean thinking can provide a useful array of tools through which the value of the design and the value delivered via the production processes can be enhanced. Sources of waste first need to be identified; and then measures can be taken to reduce or eliminate the waste, hence providing better value to the customer. This applies equally to process waste (for example disrupted work flow due to incomplete information) and material waste (for example the over specification of materials). The lean principles are to eliminate waste by:

- precisely specifying value: from the perspective of the (ultimate) customer;
- identifying processes that allow customer value to flow (the value stream) and eliminate, or mitigate, non-value adding activities;
- enabling value to flow: manage the value stream and interfaces between steps to ensure work flows uninterrupted;
- establishing the 'pull' of value: listen to the customer and do not make anything until it is needed;
- pursuing perfection by continuous improvement.

Mapping and management of flow

In the complex and dynamic project environment, the mapping and management of flow is arguably more challenging than relatively stable and repetitive manufacturing. Shingo (Shingo *et al.*, 2007) introduced the concept of flows, the flow of information and the flow of resources, as being fundamental to understanding production and to making it more efficient. These two concepts are equally relevant to construction which relies on considerable quantities of information and resources to realise a project (see Tribelsky and Sacks, 2011). Addressing flow and Ohno's (1978) seven sources of waste from the very start of projects, it is possible to bring about: improvements in quality and performance; quicker, safer, construction; reductions in whole life costs; and, less waste and better value. The seven sources of waste (Ohno, 1978) were extended to eight by Koskela (2004) with a 'making do' category for the construction sector:

- Overproduction: doing more than specified.
- Waiting: for information, decisions, materials, etc.
- Transportation: e.g. moving materials unnecessarily around the site before they are used.
- Processing: unnecessary actions.
- Inventory: unnecessary storage of materials, plant and labour.
- Movement: flow of information/people/materials.
- Making defective products: defective information and work.
- Making do: not completing a task, thus preventing the flow of work.

Interpretation of the sources of waste tends to vary depending upon one's role within a project. The important point to make is that waste should be addressed at all stages of a project and be 'designed' out. A similar argument can be made for health and safety, designing out risk and designing in safety. This requires a lean culture within all of the organisations contributing to the project. Lean culture is about doing our jobs better, engaging people and managing the process effectively: essentially a culture of continuous improvement and integrated project delivery (Mann, 2010; Santorella, 2011). Ingrained habits such as 'making do' should, and can, be eliminated by paying attention to the way in which people go about their daily duties and discussing how their work flow could be improved; adopting a lean thinking approach can make a significant contribution to process improvement (Emmitt *et al.*, 2012).

Lean design management

It may be obvious, but being able to understand and map flow, specify value and eliminate waste (inefficiencies in processes) are fundamental to the development of design management. Whether or not this is marketed under the banner 'lean design management' is largely immaterial; what is more important is that design managers incorporate these ideas and apply lean thinking at the very start of projects. Putting effort into the early stages of projects in an attempt to reduce later uncertainty is known as 'front end loading' where value is created and waste eliminated (or at least mitigated). By adopting such an approach, it is possible to significantly reduce the amount of unnecessary work on the construction site and hence contribute to a much more harmonious and professional design management process, as demonstrated by Thyssen *et al.* (2010). The philosophy of getting it right early, putting the right processes and protocols in place, is illustrated in Case study 3A; and is fundamental to value management approaches (see Case study 5A).

Case study 3A

Getting it right first time: the key to successful design management

Gerard Daws, Design Manager and Director for NBS Schumann Smith

Since the explosion in social media and networking websites such as Twitter and LinkedIn over the past couple of years, the status of the design manager as a professional role and discipline in its own right has increased. It is encouraging to think that this trajectory will increase with a bit of momentum. We are now able to communicate with each other, share experiences, challenge existing norms and provide mutual advice.

One of the most apparent developments has been the wide divergence in the definition of what a design manager actually does, across both disciplines and geographies. Admittedly, this is not a new phenomenon. A design manager working for a contractor in the US, for example, may carry out a very different role from a design manager working directly within the design team in the UK. One may be carrying out a technical role while the other may be more focused on strategic management issues and organising the design team. Although the individual roles may be different, there is a set of common themes relating to the factors for project success that run across all definitions and roles. Preparing a design programme that actually reflects how the design process will take place, ensuring that the cost management process is fully entwined with the design process and defining what will be delivered at each milestone (the level of detail is crucial, rather than just a customary list of deliverables) are all important factors for success.

Getting it right early

For our role of managing the design process within design teams for architects, one of the most important factors for project success is 'getting it right early' – putting in place the correct procedures, good habits and ways of working from the start.

Getting it right early is just as applicable for the various strands and definitions of design management as our own. The consequences of not putting into place these measures will soon be apparent across any project. Abortive work due to miscommunication, confusion over delivery dates and doubling up of roles can all be consequences of avoiding tasks that are both sensible and simple. Likewise, we have found from experience that attempting to put in place good habits and procedures once a project has commenced is a challenge as habits have already formed and cynicism about the ability to recover a situation has increased.

When is the best time to start thinking about getting it right early? We have found from experience that this commences before the project has been won. It makes sense to have a baseline set of procedures and protocols in place to be followed once the project commences. Typically design competitions request information from designers on how they will manage the team, lead the design coordination, meet the programme dates, etc. These baseline procedures can be tailored to respond to these questions but it is important that they are project-specific rather a set of standard responses.

Likewise, it makes sense that there is someone specifically responsible to lead the design management process once the project commences. From experience though, this should be preferably someone who is actually willing to take on the role rather than someone who is coerced. If team members are shoehorned into a role that they do not want to perform, it is only a matter of time before they revert back to the usual day job.

Tips and actions for success

What other tips and actions can be carried out to help implement the correct habits and procedures? There are several practical measures which can be usually carried out quickly and efficiently:

- Read the appointment. It is amazing how often this simple task is ignored by those working on project teams particularly when it is easier to regress to personal comfort zones.
- Communication is a major factor for project success. A simple project directory is a pre-requisite. Communication routes need to be established and an understanding of reporting procedures with the client.
- There can sometimes be a comforting feeling that the production of a weighty document such as a Project Execution Plan or Design Management Plan automatically means that everyone is aware of what needs to happen. In reality, even if team members actually do refer to these documents, there needs to be a constant back-up of phone calls and face-to-face meetings to reinforce the procedures and protocols in place.
- Architects with a responsibility for the appointment of the other key design disciplines under a single appointment should understand that this brings an additional set of responsibilities. Sub-consultant appointments should be back to back. Scope needs to be allocated internally within the design team and communication issues with the client need to be laid out with sub-consultants early. For example, some lead consultants are not comfortable with sub-consultants contacting the client directly under this set-up.
- Get everyone together as early as possible. We would recommend a kick-off meeting early in the design process so that key issues can be understood and discussed, such as:
  - How will the design team exchange information?
  - Is it clear what will be delivered and by when?
  - How will progress be reported?
  - Do you have all the correct disciplines in line?

- How will the client provide feedback to the design team?
- Have all statutory approvals been identified?
- Who is responsible for coordination?
- Is it clear what level of design information will be provided?
- Is documentation format clarified?
- What standards will be followed?
- Have areas of specialist design been identified?
- What information is required by third parties?

Concluding thoughts

Through experience, we have found that once the appropriate procedures and protocols have been established early enough and with enough conviction, they become second nature within the design team very quickly. The key terms here are 'early enough' and 'with conviction'. As the adoption and wider use of BIM takes hold throughout the construction sector, a new set of procedures will undoubtedly become necessary to suit the technology and the different ways of working, but the principles behind getting it right early will remain.

Greener design management

There is also an argument for better design management in respect of environmental sustainability and its links to more efficient working. The term 'sustainable development' came into common usage following publication of the *Brundlandt Report* (World Commission on Environment and Development 1987) and further attention was generated by the Rio Earth Summit conference of 1992 and the widespread adoption of *Agenda 21*. In 1997, the Kyoto conference resulted in an agreement to reduce greenhouse gas emissions by 20 per cent (based on 1990 levels) because of concerns over global warming. Governments around the world have undertaken a wide range of measures to try and improve the environmental performance of their building stock, mainly through legislation. Focus is primarily on reducing energy consumption by forcing designers and contractors to reduce the embodied energy of the building and lower its carbon emissions through ever more stringent building regulations and associated guidance. In the UK all newly built housing must be zero carbon by 2016 (DCLG, 2007) and other new buildings by 2019.

Sustainable principles

With the drive to reduce the carbon footprint of our building stock, it would be easy to take a rather narrow view of sustainability (energy reduction only) and overlook the wider picture. Cultural, economic, environmental and social aspects of sustainability need to be considered concurrently and in line with the principles of minimising waste and maximising value (Emmitt *et al.*, 2012):

- Cultural sustainability requires sensitivity to the characteristics of the local community. By recognising cultural and religious diversity it should be possible to make a positive contribution to society. This may be as subtle as engaging with the local community and incorporating local detailing traditions into new building styles.
- Economic initiatives may relate to affordability and whole life costs: the use of local materials, products and suppliers to sustain the local economy; creation of new markets and products in response to environmental legislation, etc.
- Environmental aspects include, for example: efforts to reduce waste; energy efficiency and carbon neutral buildings; improve the quality of the internal environment by eliminating toxins and improving air quality. Other initiatives relate to the use of renewable and natural materials, adaptability and the reuse of materials.
- Social aspects relate to ethical sourcing of materials and consideration for the environment and employees; the health, safety, wellbeing and comfort of workers and building users; community involvement and empowerment; and responding to the local cultural context.

Sustainable design management

Environmental sustainability needs to be built into all stages of construction projects and requires awareness by all contributors to projects and close integration between all suppliers (Rohracher, 2001; Sodagar and Fieldson, 2008). Design managers play a part here in terms of leading, promoting and often defending sustainable design ideals as the project progresses from idea to building. Here, the skills of the design manager come to play, as identified by Mills and Glass (2009) study of design managers in the UK and supported by Rekola *et al.*'s (2012) investigation of design managers in Finland.

Design managers will be involved in environmental compliance, working with tools such as BREEAM in the UK and LEED in the USA. Here the design manager has two tasks: first, to ensure compliance with the relevant legislation and codes for the project context; second, to ensure compliance is maintained throughout the project. Other responsibilities include overseeing the responsible resourcing of materials and the reduction of waste, in both material and process. Reflection on this task can be found in Case study 3B.

Case study 3B

A design coordinator's responsibility towards environmental compliance

Philip Davies, VINCI Construction UK Limited

Since graduating from Loughborough University I have undertaken the role as a design coordinator (design manager) for a major contractor, VINCI Construction UK Limited. I work within VINCI's Retail and Interiors Division which specialise in the procurement and construction of new-build and refurbishment supermarkets, retail outlets and distribution centres. Recently I have been involved in construction of two large distribution centres (one ambient and one temperature-controlled) in England for large UK grocery retail chains. Both retail chains (i.e. clients) regarded the environmental performance of their respective projects as a Key Performance Indicator (KPI) to determine project success. I was tasked with the responsibility of ensuring both projects complied with the clients' environmental expectations. Both distribution centres were granted planning permission on the condition that both upheld a strict environmental regime during design and construction. This was done by targeting respectable BREEAM scores for design and post-completion stage certification. Due to my background and overarching sustainable ambition I was asked to execute a pragmatic approach towards bridging the gap between design and environmental management throughout the development of both projects. It is this experience that informs this case study.

In my experience, emphasis on reducing the environmental impact of construction has been truly reflected within both academia and industry. A new generation of pioneers are being nurtured to consider the practicalities of implementing a sustainable agenda from early in the life of projects. Contractors such as VINCI have an opportunity to capture the innovative (untainted) views of young professionals, who have yet to be influenced by the commercial boundaries that underpin decision making. It could be argued that altering the existing construction culture appears to be more of a challenge compared to bringing in fresh ideas from the new generation of graduates, who are keen to enhance attitudes and develop new, greener, ways of working. Clients also have a role to play, with clients' environmental obligations becoming ever more demanding; and hence more challenging to achieve from both a technical and procedural perspective. These changes also have to be seen in context of increasing legislation and pressure from central and local government for a more sustainable built environment. But contractors are also required to react instinctively towards changes in client demands and understand the full ramifications of the changes in respect to design, construction and environmental performance.

My role

As a design coordinator, my role primarily entails capturing, disseminating and re-directing information in a suitable format for those who need to use it, commonly via a collaborative working package such as 4Projects extranet. My role helps to prevent (or remove) potential contractual disputes while also positively influencing the management of activities to reflect harmonisation of information across all project stakeholders (client, contractor, sub-contractors and consultants, etc.). Fundamentally, my role centres on questioning all of the information provided to the contractor. This involves solving problems which predominately stem from the design

Figure 3.1 Going about my job on the site
Photographs courtesy of Stephen Emmitt

development stages – specifically the lack of legitimate, detailed, information – which prevents effective coordination and critical analysis of design information. As a result, I would summarise my role as being reactive, constantly responding to unforeseen circumstances to try and prevent uncertainty and its immeasurable consequences for the project. For example, my immediate challenge is undertaking decisions which do not adversely affect the management of the four core key performance indicators (KPIs) which, for decades have determined project success within construction and wider industries, namely: time, cost, quality and (more recently) health and safety performance. I believe there is a fifth core KPI – environmental management – which should be thoroughly considered during project decision making; although from my experience this is not generally applied in practice.

It appears to me that every decision one makes in construction (from procurement to on-site activities) has either a positive or negative effect on the five core KPIs. Because environmental management is 'new to the table' this KPI is commonly ranked the lowest and consequently usually receives the least consideration. Moreover, similar to good health and safety management, successful environmental management does not come cheap and requires extensive deliberation throughout all stages of project development to apply it successfully. Here lies the challenge. As in my project examples, if project environmental performance is important to a client, then the significance of environmental management and the parameters which determine environmental compliance need to be clearly expressed and agreed by all project stakeholders at the outset. In practice, this is very difficult to achieve because although the contractor is contractually responsible for delivering the targeted BREEAM rating, it is the consultants (e.g. architects and engineers) and sub-contractors who are the primary providers of design information, with their activities significantly impacting BREEAM performance.

A similar view can be exhorted on the management of material waste on the construction site. Unlike BREEAM, the Site Waste Management Plan (SWMP) is a joint legislative requirement between the client and contractor, whereby VINCI is responsible for ensuring that construction waste is managed in the most legitimate, environmentally sustainable, way possible. The performance of the SWMP is idealised on the activities and decisions undertaken by the consultants and sub-contractors during design development and on-site operations. Again, my role tends to be reactive, dealing with unforeseen circumstances as they arise. Table 3.1 illustrates my overall core roles and responsibilities as a VINCI Design Coordinator.

Total project environmental performance

Principal contractors such as VINCI pride themselves on their environmental performance, as reflected within the organisation's corporate mission statement. However, in practice it can be difficult to ensure that all aspects of the

Table 3.1 Design coordinator roles and responsibilities

Function name	Role level	Responsibilities
4Projects (4P) extranet	*Management*	– Print and forward drawings to client weekly – Organise, construct and maintain approval routes for consultants and sub-contractors – Capture drawing comments from consultants and sub-contractors – Undertake approvals of drawings – Upload drawing comments and notify associated consultants and sub-contractors – Write and publicise best practice guidance – Maintain approved drawing register and folder.
Best practice procedures	*Assistant*	– Ensure project team adheres to best practice in terms of on-site procedures – Review current procedures and identify areas of improvement – Report findings back to project team.
BREEAM assessment	*Management*	– Arrange and chair meetings – Deliver minutes and actions – Capture required information from consultants and sub-contractors – Provide feedback to consultants, client and project team – Micro-manage BREEAM assessors.
Community involvement	*Management*	– Arrange community and educational site visits – Provide and deliver supplementary information and presentation material – Record visits and document evidence for internal and client purposes – Evaluate visits and identify areas of improvement.
Considerate constructors scheme (CCS)	*Management*	– Arrange and attend CCS audit meetings 1 and 2 – Capture required information from consultants and sub-contractors – Produce information pack – Encourage considerate behaviour on-site.
Environmental board	*Management*	– Transfer required information from BREEAM, SWMP, EPI and publicise on-site – Display monthly fuel consumption, water consumption, and site delivery energy consumption data.
Environmental performance indicators (EPI)	*Management*	– Capture required information from sub-contractors – Ensure sub-contractors are legally compliant – Arrange and chair pre-start meetings – Write and publicise best practice guidance – Produce monthly progress reports to project team, client and internal organisation – Encourage good energy management practice on-site.

Table 3.1 continued

Function name	Role level	Responsibilities
Internal environmental audit	*Assistant*	– Organise and attend audit site visits – Collect necessary environmental data before hand and display on the environmental board – Demonstrate environmental compliance, procedures and systems during audit – Review previous audit score with project team and identify improvements.
Material samples	*Assistant*	– Obtain material samples from manufactures and suppliers – Obtain formal approval from consultants and client team regarding selection of material.
Operational and maintenance manual (O+M)	*Assistant*	– Capture required information from sub-contractors and consultants – Formalise information with project team and gain approval.
Project directory	*Assistant*	– Capture and record organisational information from sub-contractors and consultants – Issue updated registers to project team.
Quality assurance (QA) documentation	*Assistant*	– Capture required information from sub-contractors and consultants – Formalise information with project team and gain approval.
Site inductions	*Assistant*	– Provide content and aid the structure of the induction – Continually update the induction to demonstrate project progression and associated on-site risks – Provide inductions to sub-contractors when required.
Site logistics	*Assistant*	– Maintain and circulate plans periodically across the project team – Review content of plans with project team and make necessary changes.
Site waste management plan (SWMP)	*Management*	– Capture required information from sub-contractors and waste contractors – Ensure waste contractors are legally compliant – Input data in to the SWMP – Write and publicise best practice guidance – Produce monthly progress reports to project team and client – Encourage good waste management practice on-site.
Sub-contractor proposals	*Assistant*	– Comment on sub-contractors' proposals in terms of specification and compliance – Comment on suitable alternate proposals for specified materials.

Table 3.1 continued

Function name	Role level	Responsibilities
Technical queries (TQ)/ Request for information (RFI)	*Management*	– Ensure all incoming/outgoing information is captured on the register – Capture required information from consultants and sub-contractors – Issue outstanding requests when required – Close out unresolved issues.
Validation and remediation report	*Management*	– Arrange and chair meetings – Deliver minutes and actions – Capture required information from consultants and sub-contractors – Provide feedback to consultants, client and project team – Undertake on-site measurements of soil depths, quality and document.
Value management risk and opportunities register	*Assistant*	– Develop initial risk and opportunities registers – Ensure all developments during VM meetings are accurately recorded on the registers – Distribute the registers to all team members and chase up outstanding actions.

corporate mission statement are strictly obeyed throughout project development. Complications sometimes arise due to confusion surrounding the plethora of internal (organisation) and external (client and wider industry) policies, procedures and systems that are required nowadays, accompanied by a 'never ending' trail of information to ensure full environmental compliance. If the true importance of specific environmental expectations are not clearly understood and frequently reiterated by project team members to consultants and sub-contractors, I begin my task of ensuring project environmental compliance on the back foot, which is not an efficient way of working.

Occasionally decisions are made that are beyond my comprehension, which are expected to be in the best interests for the project and client, but which cause difficulties in maintaining compliance with my organisation's internal policies. In such situations I am responsible for capturing, analysing and reporting project environmental performance both internally and externally. It is also my obligation to act as an 'internal consultant' and convey the unforeseen consequences of decisions to senior project management to ensure the overall best decision is undertaken. For example, changing the approach towards managing waste from contractor to sub-contractor responsibility has obvious (and some not so obvious) consequences. Direct commercial benefits can be achieved for the contractor (by specifying fewer skips on site, etc.), although this can be at the expense of losing direct control of the requirements which govern legal waste management practice, hence creating direct challenges for the site management personnel and indirect challenges for the design managers/coordinators.

Waste management is commonly perceived within contracting organisations as a construction management issue and not a design management issue. However, as waste management performance is a legislative and contractual requirement the importance of waste management is also reflected within BREEAM; which design coordinators are usually responsible for. Therefore, in the best interests of total project environmental performance, and as a design coordinator who is passionate about environmental performance, I took it on myself to become involved in the full scope of waste management, thus helping to benefit the project as a whole and also my organisation. In doing so it became evident that the design coordinator plays an important role as both mediator and informer. Not all stakeholders who make decisions are responsible for data processing or the audit trail which is required by both client and contractor, thus I have to be proactive in taking a lead on environmental compliance. I have found that taking a proactive approach helps to save time for all stakeholders and leads to a smoother, more efficient, process.

As a design coordinator I am expected to give the construction management team the confidence that all potential difficulties surrounding installation (buildability) of building materials and components have been identified and solved before on-site activity commences. If problems arise which have not been previously identified (e.g. clash of drainage connections, interface issues between cladding and glazing panels) it is difficult to fast-track a solution which can be agreed and is in the best interests of all project stakeholders. Construction programmes allow a very small 'window of opportunity' to get building materials and components installed before another sub-contractor begins work. Thus unforeseen problems on-site generally have a detrimental impact on the five core KPIs, often at the expense of the contractor. Therefore, I am expected to be aware of the full implications of every decision undertaken and how this could benefit my employer.

Overall, the best aspect of my role as a design coordinator is the opportunity to engage all levels of stakeholders during design development and on-site activities, in order to enhance project design and construction. Regardless of the impact of my role towards other stakeholders, my primary focus is to ensure VINCI is aware of, and can potentially benefit from, all decisions undertaken regarding the five core KPIs, while still delivering on our contractual obligations.

Future directions

In the future, I expect environmental management to play a more influential part in construction projects than it does at present. Design managers and design coordinators are currently expected to spearhead environmental obligations and commitment. We have a unique opportunity to be involved in all forms of interaction between project stakeholders with the prospect of adding value and ensuring environmental compliance is adhered to throughout project development. Stakeholders need to become more aware of the consequences of their decisions and their impact on the project's five core KPIs. Information requirements need to be addressed early during development in order for all information to be in a clear, consistent, and agreed format which can be easily disseminated by all stakeholders in order to positively influence informed decision making. Advances in ICT, 4Projects extranet and BIM (i.e. collaborative working packages) will be helpful in this regard, but there is still a need for design coordinators and design managers to take a lead and be proactive if we are to achieve a more sustainable built environment. As advances in BIM and future reliance on a collaborative way of working become apparent, I believe the role of the design coordinator and design manager will evolve in order to lead and manage the integration of BIM within project development, thus moving from a reactive to a proactive role in the delivery of design value.

Conclusion

Design management as a discipline has been around since the 1960s, although it is only more recently that the role has been taken up with enthusiasm by members of the construction sector. As demonstrated in the literature and within the case studies the role is quite diverse, depending on whether the design manager is working for a design practice or a contractor. However, some of the fundamental, underlying, characteristics associated with the role are starting to emerge which relate to:

- process improvement;
- demonstrating the value of design;
- early involvement;
- integration.

In Chapter 4, attention turns to the design manager's role from the perspective of the contractor, with a number of case studies helping to demonstrate the scope and richness of the role.

Further reading

For an overview of how to put committed teams together within a lean thinking ethos, see Gary Santorella's book *Lean Culture for the Construction Industry: building responsible and committed project teams* (CRC Press, 2011). Additional information on the generic design manager role can be found in Brigette Borja de Mozota's *Design Management: using design to build brand value and corporate innovation* (Allworth Press, 2003).

4

THE CONSTRUCTION DESIGN MANAGER

In many respects, the expansion of the construction design manager role reflects the constructors' growing awareness of the value of design, changing procurement routes, evolving information communication technologies, increased responsibility for design quality. And also recognition of the benefit of managing the design at all stages of construction; which is true of the new build sector and the refurbishment sector. Job vacancies for design managers are being advertised as 'pre-construction design managers' or 'construction design managers' to fulfil different, yet complimentary, roles. In this chapter the emphasis is on the tasks and responsibilities of construction design managers at the key stages of projects, be it pre-construction, construction or post-construction; illustrated within this chapter by three case studies from practitioners.

The contractor's perspective

As the design management discipline has matured, the role and responsibilities of the construction design manager have expanded. Initially design managers were located on the construction site, dealing primarily with issues concerning the coordination of design information (including clash detection), requests for information (RFIs) and managing design changes. More recently, construction design managers have become involved with pre-construction design activities, ranging from; first contact with clients, client presentations, dealing with town and country planning applications, coordinating and managing the development of the design, dealing with sustainability issues and environmental compliance (using evaluation tools such as BREEAM and LEED) and reviewing the design to ensure that health and safety risks are designed out or mitigated (e.g. complying with the CDM Regulations). These pre-construction design tasks would traditionally have been associated with architects, engineers, and designers, but they now come under the remit of contractors as they take on greater responsibility for design. The role of the construction design manager is also changing as building technologies change and information technologies become more powerful. Advances in production and the availability of off-site technologies (modular and prefabricated components), combined with a greater awareness of environmental issues and more stringent legislation relating to low carbon building design help to emphasise the importance of design and its management. With the adoption of BIM, the role of the design manager is also being pushed towards earlier involvement in the design process, since it is here that many of the problems of coordination, clash detection and compliance are addressed and resolved collaboratively in a single virtual model before the physical act of construction starts.

Construction design managers are employed to lead design-related tasks and are usually responsible for the design teams contributing to the project. These teams may be internal to the business, for example a major contractor's in-house design team, or they may be external to the business, employed to provide design expertise for a specific project.

As noted in Chapter 3, outside the AEC sector the design management roles may be related to areas such as product development, advertising and web design, working in a variety of industries, from manufacturing to the creative arts and advertising. Within the AEC sector the design manager role tends to be associated primarily with large to medium-sized contracting organisations, although the role is enacted within design and engineering organisations, often via a different job title (such as 'project architect'). The design manager role is usually project specific, with individuals tasked to manage all aspects of design on one or more projects. This will involve interaction and collaborative working with other departments and external organisations.

Emphasis on sustainability, especially cultural sustainability, along with economic concerns and restrictions on developing green field sites has resulted in an increase in work to existing buildings. This is important in the drive to make the existing building stock more energy efficient, while contributing to the existing character of our built and cultural heritage. In the refurbishment, fit-out, and retrofitting sector, the emphasis is on the coordination of historical design information with newly generated design information to ensure harmony between old and new. Existing drawings are rarely an accurate representation of the building, emphasising the need for measured (and building condition) survey work. Compared to new build schemes, work to existing buildings represents increased risk because of uncertainties relating to the condition of the building and the materials used (e.g. asbestos) and the cost of dealing with little 'surprises' as the building fabric is opened up. This calls for individuals with different knowledge and skills sets, resulting in construction design managers who specialise in new-build projects and those who specialise in refurbishment and heritage projects. For projects that comprise both new build and refurbishment work, it is necessary to employ design managers from both areas of specialism to ensure design value is delivered to the client in an effective and efficient manner.

Education and training

Most contracting organisations have some form of career structure for design managers, related to experience and responsibilities. Given the relative newness of the discipline, it is common for contractors to allow both a vocational and an academic route for career development. This usually takes the form of a hierarchy, which moves from managing documents, to managing the process, to managing people and providing design leadership across multiple projects at the top:

- Document control. This is the entry level, with individuals holding an HNC or HND. This is where aspiring design managers learn about design management and also get to understand how the business is managed by working with design documentation. Once they have developed an appropriate level of understanding, they can progress to a design coordinator role.
- Design coordinator (trainee design manager). Design coordinators will need at least a first degree in a building discipline that includes design education, such as architecture, architectural engineering, architectural technology, building surveying, design management, civil engineering, services engineering or structural engineering. The three case studies in this chapter are written by graduates of the AEDM programme at Loughborough University. Alternatively design coordinators may have acquired the appropriate practical experience to undertake this role, perhaps supported with CPD activities. Design coordinators are usually tasked with managing the process.
- Design manager. Individuals will have qualifications equivalent to chartership with the CIOB. It is at this level that leadership abilities start to emerge, people skills being developed and honed and commercial sensitivity developed by working on bid management. Design managers are usually tasked with managing people and processes and an important skill will be the ability to work across institutional and disciplinary boundaries.
- Design director (senior design manager). The design director will work across multiple projects on behalf of the contracting organisation. He or she will form an important interface between the business objectives of the organisation and the organisation's project portfolio. This position demands excellent leadership skills, people skills and commercial management skills.

Progression from the bottom to the top will depend as much on experiential learning as it will on qualifications. However, there is a need for CPD and many design coordinators and design managers will need to top up their qualifications as their career progresses. Educational programmes should aim to educate future design managers to appreciate the value of design and the benefits of collaborative (interdisciplinary and cross functional) working. There is also an opportunity for universities to offer stand alone modules for design managers from a vocational background to take part-time or on a distance learning basis.

In addition to having gained experience in their own discipline, it is vital that design managers understand the business aspects of projects and their organisations in order to be effective in their role, i.e. they need to develop commercial awareness. For construction design managers, it is essential that individuals understand contracting, i.e. the importance of estimating, commercial management and health and safety as important business drivers.

One of the findings of research into improving design management techniques in construction (Bibby, 2003) was the need for better education and training and the lack of leadership from senior management. This led to the development of a design management handbook for the industrial sponsor of the research (Bibby, 2003). While the challenges identified by Bibby have not gone away, since 2003 there have been considerable advances in the education and training of design managers within contracting organisations. These tend to be a careful mix of generic leadership skills and more specific training for the role of the design manager. Many contracting organisations now run highly effective in-house training programmes for design managers, as demonstrated in the case studies in this chapter. Others, often with limited resources, still tend to rely on design managers learning on the job while working alongside more experienced colleagues, supported with a variety of CPD events.

Technical and social skills

Farr (1966) makes an excellent point that '...the design manager needs knowledge, specific working methods and skills. His [or her] tasks lie in problem solving, planning, briefing, communications and coordination. They occur within every project, but their content is never the same.' Boyle (2003) makes a similar point, emphasising the need to understand, coordinate and synthesise a wide range of inputs from a diverse range of project stakeholders. This means that the construction design manager needs to be flexible in both approach and attitude, able and willing to respond to different project contexts and work with a wide variety of project contributors. It also means that there will be tension from conflicting forces in every project, and it is the design manager's job to anticipate and resolve such tensions (Farr, 1966).

In addition to having creative flair and knowledge of design techniques, the construction design management role also requires individuals to demonstrate excellent communication skills and the ability to multi-task within a multi-disciplinary temporary project organisation. This requires individuals to be relatively agile and have the flexibility to respond to rapidly changing conditions, while remaining reassuringly consistent and dependable in their day-to-day actions. The people skills, such as diplomacy, negotiating, coordinating, communicating, integrating, and organising are central to the design management role. Although many of the skills required are similar to those required of project managers, there is a deliberate bias toward design knowledge and a deep understanding of the design process. Design managers must be able to champion design quality and communicate the benefits of good design to a variety of project stakeholders. Design managers must also possess a comprehensive understanding of building technologies (buildability), health and safety legislation, environmental legislation, and life cycle costs.

Design managers should be able to:

- champion design quality at all stages of projects;
- communicate effectively using written, oral and graphic media;

- consistently manage the production and realisation of high quality designs;
- coordinate diverse design works packages;
- lead and supervise design teams;
- listen to the concerns of their team members;
- motivate design team members;
- realise business objectives through effective design management;
- resolve design related issues and problems in a timely and efficient manner;
- work effectively in a multi-disciplinary temporary project environment.

Roles and responsibilities

Gray and Hughes (2001) have emphasised the importance of clearly identifying and communicating the roles and responsibilities of the design manager. Of particular concern is the relationship between the construction design manager and the project manager, given that many of the design manager's functions are grounded in project management.

There are some typical (generic) responsibilities that apply to construction design managers, regardless of who they are employed by or at what stage of a project they specialise in. Construction design managers are responsible for ensuring design quality is realised within the constraints of time, budget and resources. Typical responsibilities may include some of the following:

- achieving design quality targets;
- arranging, coordinating, attending and chairing meetings and workshops at organisational and project levels;
- collaborating with a wide range of project stakeholders and other departments;
- complying with health and safety legislation (e.g. CDM Regulations) and environmental legislation (e.g. BREEAM, Code for Sustainable Homes);
- conducting design reviews and design appraisals;
- creating innovative designs;
- delegating and reviewing design team tasks;
- ensuring design parameters are adhered to;
- guiding and leading the design team;
- meeting client (and stakeholder) expectations;
- motivating a wide range of designers;
- presenting design proposals;
- reviewing budgets and financial reporting;
- value engineering the design.

Some of these characteristics are further developed in the case studies, starting with reflection on the role of a graduate design manager in Case study 4A.

Case study 4A

A graduate design manager's perspective

Rachel Bowen-Price, Morgan Sindall

I work as a graduate design manager for Morgan Sindall, based in their Ipswich office. Morgan Sindall is a UK construction, infrastructure and design business with a network of local offices. The company works for private and public sector customers on projects from £50,000 to over £500 million. In this case study I have provided a perspective based on my experience, reflecting on roles and responsibilities and also the future directions for construction design managers.

Responsibilities and accountability

The Morgan Sindall design manager job role covers the following:

- manage the bid and design process to meet the agreed programme;
- assess customer documentation and requirements;
- contribute to development of tender strategy and continuously monitor design and build projects;
- consider opportunity and scope for value engineering where appropriate;
- ensure high standards of health and safety, quality and client satisfaction;
- inspect and audit design and build systems and procedures and ensure any subsequent actions required are carried out;
- maintain awareness of current Morgan Sindall health and safety requirements and changes;
- implement environmental procedures with respect to design;
- ensure effective communication to all parties;
- identify alternative materials and products;
- seek opportunities to learn about changes and innovations in the industry;
- influence and support teams to innovate and achieve optimum solutions;
- offer solutions to site-related difficulties;
- promote and present a professional Morgan Sindall image to the customer in all dealings, to become the main point of contact on design and build jobs through the bid period until start on site;
- ensure the customers' feedback and comments are acknowledged, prioritised and actioned;
- work closely with whole of design team and suppliers offering the contractors' design portion (CDP);
- continually review suppliers' performance and share information;
- influence and support suppliers to innovate and achieve optimum solutions

- deputise for bid manager when not available, during tender period;
- involvement with pre-construction including tender finalisation, tender strategy and adjudication process;
- identify new business opportunities/leads;
- facilitate and lead design management meetings and workshops.

Projects

Since joining Morgan Sindall I have worked on a wide range of projects and have enjoyed my ever increasing responsibility for design management. During the summer of 2008, I worked on the National Construction College Redevelopment, Bircham Newton. This was a tender project for the redevelopment of an existing construction skills college to create a 'BREEAM Excellent' multi-residential teaching environment, social space and student accommodation on a redundant RAF site. I assisted the preconstruction team in compiling tender submissions, working alongside the lead designer manager, planner, estimator, and business development team.

During my industrial placement year (2009–2010), I worked on Basildon Sporting Village, Basildon, Essex. This comprised a £35m sporting facility featuring: a 50m swimming pool, adjustable boom and 25m raised integrated platform accessories, stadia for 1,400 people, a gymnastics arena with jump pits, a multi-purpose sports hall (basketball, cricket, tennis, badminton, netball, fitness suite), five synthetic football pitches and a car park. This project had a 'BREEAM Very Good Bespoke' rating. As an assistant design manager, I was responsible for managing and maintaining 4Projects extranet, and ensuring designers and the site operatives had the relevant up-to-date information and sub-contractor information. Following organised drawing review meetings, I was responsible for relaying comments via our electronic mark-up tool to the relevant individuals to amend. The organisation and management of BREEAM meetings with consultants was an additional responsibility within this period of work experience, ensuring credits were being achieved at the appropriate time within the programme. Other tasks included: submitting design package information for building control approval; issuing 'client design approval information' to the client and representatives for approval; monitoring requests for information via the 4Projects extranet. Additional responsibilities included updating a monthly directors' report on design development, which included items such as: drawings issued, drawings overdue, drawings approved, request for information issued and closed/outstanding, general design consultant performance review; and managing any design related risks.

Since graduating from Loughborough University in the summer of 2011, I have been working on Bircham Newton Phase 1 redevelopment. This is a design and build contract with a contract value of £4.5million. The development includes two student accommodation blocks housing a total of 78 students; with the potential for five future development phases. The accommodation includes prefabricated bathroom pods and woodchip biomass boilers for domestic heating and hot water, to name but a few of its features. This project was reduced to an overall BREEAM rating of 'Very Good' following a value engineering exercise.

My personal role and responsibilities are:

- compiling contractors' proposals, design statements and managing design consultants in the preconstruction phase to reflect the scope of works requested in the employer requirements;
- managing the design information release schedule to fall in line with the construction and procurement programme;
- managing consultants to assist in closing out and submitting town planning conditions;
- assisting in the completion of consultants' appointment documents, including design responsibility schedules, scope of works and fee schedules;
- managing the design information release programme to match that of the as-built and current construction programme;
- raising design change proposals;
- assisting in compiling information for the client site meetings and any client request for information (RFIs);
- visits to manufacturers to ensure design and quality control of off-site production of products and components;
- management of design consultants to contribute and submit evidence for BREEAM Advisors to review;
- appointing specialist design consultants for acoustic and ecological elements of design; managing their scope of works in line with the construction programme for testing and site visits for validation;
- organising and chairing design team meetings and design workshops;
- submitting design information for building control approval (and comment);
- drawing review with the internal project team for cost, employer requirements and buildability, and consultant comments for coordination and technical approval ready for construction through the use of 4Projects 3G mark-up;
- managing the 4Projects extranet for drawing submission, managing request for information (RFIs);
- general day-to-day design development and risk management, assisting with on-site remediation and consultation with designers for solutions.

Education and training

Morgan Sindall enrols graduates on a two year graduate scheme which covers two interrelated areas, personal development and business. The personal development training programme includes topics such as; making an impact, time management, presentation skills and influencing others. Business related development is presented by key players within the company to best relay the roles and responsibilities of varied business processes. Typical topics include: supply chain management, site records, setting out, planning and programming and design appreciation. Health and safety is a major consideration. The 100 per cent Safe training covers the company's five pillars of safety: safe places of work, safe by choice, safe relationships, safe by design and safe lives. Design-specific training is occurs as a result of quarterly reviews and organised throughout the year. Training includes: environmental awareness, CDM 2007, and asbestos awareness.

Graduates are encouraged to become an incorporated member of their particular chartership. They are provided with a mentor whom they meet on a monthly basis to review progress and they have direct contact with their mentors for any queries when compiling professional evidence. Employees are also encouraged to attend CPD events and internal lunchtime events are regularly held for people to attend.

What I like about my job

There are many aspects I like about my job. I enjoy the mobility that goes with the role: moving between various locations for different projects and from the site to the office and vice versa to do my work. There are also many opportunities to visit manufacturing plants, see different products develop and learning about the varied manufacturing processes during my visits. Being able to see the complete design develop out of the ground from substructure to completion is always rewarding; as is working as a team and within a multi-disciplined team. It is this interaction with many specialists that aids my learning, new skills development and general knowledge. In particular, being able to learn more about the client and the management (function) of the building is rewarding, as is dealing with day-to-day changes and responding to the challenges faced.

Room for improvement

There are a number of factors that could be improved. The uptake of collaborative extranets such as BIM and 4Projects could help to improve collaborative working, both amongst project team members and external consultants. Related to this is the slow uptake of innovative communication technologies for managing processes and managing communication, which is currently on the market and used effectively in other business sectors (such as interactive whiteboards, ipads, etc.). Faster adoption of technologies may help to improve collaborative design management and also help design managers to be more efficient and effective in their job. Finally, IT installation of start-up construction sites is very slow and restricts the speed at which people can work on site when visiting or while trying to promote collaborative technology and ICT.

Future directions

As the next generation of built environment employees with experience and exposure to information communication technologies enter the workplace, the progressive acceptance of its use in the construction sector will improve. Lack of experience has raised initial ambiguity toward new tools such as BIM, however, over time I believe confidence due to experience, increased levels of successful precedent studies along with government plans to restrict contractors' involvement in government-funded projects unless they use BIM, will encourage future collaborative tools. Furthermore, with generations becoming more experienced in technology, this will ease the strain in dividing opinions towards the likes of extranets. Collaborative tools shall become more common and the norm instead of a side process, where collaborative tools are currently an 'add on' to the traditional ways of working which still reign.

Pre-construction – generating design value

It is during the design phase that value is created for the construction client, and this phase must be managed effectively to ensure that maximum value is created and waste eliminated. It makes sense therefore for contractors to address design issues as early as possible. Indeed, failure to manage design during this phase of the project may result in additional, arguably unnecessary, work during the construction phase as contractors seek to change decisions made earlier in the process.

Similar to site-based activities, the design manager's role varies considerably at the pre-contract stage, dependent on the market orientation of the contractor and the type of procurement route being used on a specific project. As a broad generalisation, the pre-construction design manager's role is to manage design information and facilitate effective design, with a view to maximising the value of design. His or her role is not to create the design or produce design information; that is the role of the design team, such as the architects, architectural engineers and technologists, services engineers, and structural engineers. Working alongside the contractor's project manager the pre-construction design manager must ensure that an appropriate design management framework is put in place to facilitate the efficient development of the design, i.e. a framework that allows all contributors to interact effectively. This means that some pre-construction design managers will be responsible for assembling the design team and allocating roles and responsibilities. Some will also be responsible for the client interface, managing the briefing stage and establishing the employer's requirements.

Job advertisements for pre-construction design managers ask for previous experience of working for (major) contractors as an essential requirement. They will also ask for experience in the coordination and/or management of design and the ability to take the project from inception through to the detailed design stage. This includes experience of tendering and bid procedures. The role is usually advertised to include the following tasks:

- assess design details and design information for buildability;
- attend pre-start and progress meetings;
- conduct value engineering exercises and implement innovative solutions;
- coordinate internal and external design consultants;
- develop and manage client proposals, including pre-qualifications and approved lists;
- ensure bid proposals align with client requirements and budget (work with account managers and bid managers);
- liaise with health and safety managers and environmental managers to ensure designs comply with relevant requirements, such as the CDM Regulations, BREEAM and Code for Sustainable Homes;
- manage information flow (provided on time);
- manage the client interface;
- manage the approval process (town planning, highways, building control, etc.);
- monitor design costs associated with the building process;
- support the bid manager and liaise with the tender bid team;
- take a lead role at tender stage for the design process;
- undertake site (land) reviews.

Relating this to the RIBA Plan of Work, the main stages to be managed range from Stage A 'Appraisal' through to stage H 'Tender Action'. One of the early stages, development planning and feasibility studies, is explored in more depth in Case study 4B.

Case study 4B

Development planning and feasibility

Tom Warden, Tesco PLC – Property

I work as a development planning manager for Tesco Property within the architecture and programme function. My role is to investigate the feasibility of potential store developments, either as new space or the extension of an existing store. The role provides a significant commercial benefit to my company, as I work up the best possible design under the site constraints and gain the cost and expected sales to identify if the scheme will be profitable for the business. For this reason, the majority of schemes I work on are proven not to be profitable, and as such are abandoned before any money has been spent by our company. Approximately 1 in 20 schemes that I work on will go forward and make its way onto a development programme, which can be up to eight years in the future. By employing design management at this early feasibility stage, we are able to gain best value from our properties.

I employ design management in my role on a daily basis due to the nature of the work. Developing a scheme requires input from a variety of both internal and external sources. Process driven management of this information and the people involved is essential to getting best value from the project. Information about the site is compiled by several departments within Tesco to establish what we need to deliver such as sales floor area, the number of parking spaces and any easements or planning constraints that may exist on the site, as well as area affluence and the presence of any competitors in the locality. I then take this information and communicate it to various consultants such as architects, highway consultants and retail planners to develop a site and retail plan that is sympathetic to the site and delivers the needs of our customers. The drawings produced can then be costed by our value managers to identify whether the design provides value against the investment.

My remit within development planning is feasibility of our Metro format stores. This is one of our smaller format stores that is usually found in town centres as a convenience store. Many more challenges are present when designing this format of store, because the constraints posed by dense town centres impose more challenges for the design. By addressing each store at the feasibility stage, we deploy more resources into the design of our buildings so that consideration of various stakeholders can be carefully captured and included in the design. Typical issues I address in Metro feasibility are elevation design due to planning constraints on urban context, irregular column grids that compromise retail planning and design of the back of house areas and deliveries, which are often located in other levels of the building.

A key skill required in my role, which is vital to effective design management, is managing communication. I have to liaise with many internal and external consultants to gain information and commitment to develop the scheme. For this reason, good people skills are needed to influence and gain commitment. Effective time management is also required to understand the time taken for each consultant to perform their work so that resources can be allocated in the correct sequence and the whole process can operate as quickly and efficiently as possible. I then assess the output of each consultant and instruct changes that need to be made to the design either based on the information gained from varying stakeholders or through experience of similar successful schemes we have delivered in the past.

Tesco is different from other organisations in the built environment because we sit on both the client and delivery sides of the fence. A heavy focus is put on good design and value creation in our projects because as the client, we directly feel the impact of that building while in use through the sales achieved in the store. To improve our sales as a company, we actively need to deliver what our stakeholders require; our customers and staff being the majority stakeholder in everything we design and build. For this reason, another important part of my role is visiting operating stores across our estate and understanding how successful their retail operation has been as a result of the store design. This allows me to capture best practice and look to reuse those elements in my work to create continuous improvement to our new stores.

Projects

A recent scheme I have worked on was the extension of one of our existing superstores. The driver for extending this store was to increase footage to house a better range of products for our customers. Due to several competitors opening stores close to our site, failure to improve our food and non-food offer could result in losing customers and sales to other businesses. The scheme had a number of challenging factors to deal with. The extension would increase our floor area by 25,000 sq ft, changing the store from a Superstore format to a Tesco Extra format, which would completely change the internal layout of the store and the range we provide. Likewise, the increase in trade would add pressure to our access and egress and car parking. Due to the increase in competition in the locality, the store's car park became very busy and we felt that customers of other retailers were abusing our free parking policy. The solution to this was discussed in the feasibility design stage, with the solution being to include a registration plate recognition camera system to the store's car park.

Historical information of the site identified an ancient burial ground very close to the site. This meant that particular resources needed to be invested in geotechnical surveys to identify the location and extent of the burial ground. Careful consultation and exchange of information between architects, highway consultants, geotechnical engineers, site researchers and retail planners meant that we were able to identify the true value of the proposed extension before any capital was employed to extend the store. By involving all the stakeholders at the concept stage of the design process, greater value was achieved as many costs and poor design decisions were mitigated. The economic climate would suggest that a high-risk scheme such as this would be unlikely to go forward, as the sales generated against the cost of the development would not be a safe enough margin. The work I performed for this scheme identified this risk and saved the company from spending money that would have otherwise been unforeseen, had this careful work at the design phase not been carried out.

Many of the consultants we use in Tesco are from approved lists that we have worked with on many projects in the past. Due to the volume of work we undertake in Tesco Property, it is essential that our consultants understand the way we work and we understand their methods as well. This has led to a more collaborative approach to working which has seen a massive benefit in our projects. Strong working relationships have been established and time savings have been felt as many of the preliminaries are already in place. A number of our consultants have worked in the office to integrate them to our teams for a fixed term to improve relationships and develop methods of working between the bodies. I believe this is something that

Tesco will continue to do with many of our partnered consultants as the benefits have been substantial. Tesco Property also hosts a number of projects and competitions within the industry to stretch our existing supplier/consultant base and find opportunities to identify new partners to work with. Taking this collaborative approach to design and build of our stores has made the process very efficient and delivered better value from our developments.

Education and training

Training at Tesco Property is focused around personal development through the use of its leadership framework. This is divided into three areas: Improving the Business for Customers; Taking People with You; and Living the Values. All three areas are aimed at improving management skills through relationships, rather than process, to promote macro-management. As a design manager, these skills are integral to providing best value and output from the design team. I have attended courses such as Momentum and Value Creation, Analysis and Decision Making, and Personal Management, all of which provide tools for increasing efficiency, but focusing on team working and relationship building.

As a member of AEC, it is important to remain educated in the latest developments in the built environment. I am encouraged by my company to attend professional conferences, seminars and events that affect the business, to develop my industry knowledge and to drive innovation, and hence remain competitive in the industry. Tesco Property also supports a number of professional qualifications such as RICS, CIBSE and CIOB, the latter of which I am currently completing. On successful completion of the programme, I will take on the role of a CIOB mentor for prospective CIOB applicants and guide them through the chartership process.

Construction – delivering design value

Regardless of the type of procurement route used, it should be recognised that there is a change of culture as the project moves from the creative design phase to the realisation phase. The design team should have completed their task, or the majority of their work, to enable the constructors to realise the design. Design intent will have been codified in drawings, schedules, written specifications and models, both physical and virtual. The task of the constructor is to interpret the design information and translate it effectively and safely into a physical artefact. Here the construction design manager plays an important role, spanning the boundary between design and production, helping to resolve queries and assist with interpretation of information to facilitate buildability. The primary task is managing information: coordinating information from many different sources and managing the flow of information to ensure that delays do not occur on site because of missing, incomplete or erroneous information. Emphasis is primarily on ensuring that design value is delivered to the client. Changes and value engineering exercises should aim to improve design value, not compromise it. Thus construction design managers will span the design and construction phases, as demonstrated in Case study 4C.

Similar to the pre-design management position, the job advertisements for construction design managers ask for previous experience of working for (major) contractors as an essential requirement. They also ask for experience in the coordination and/or management of design (information) and the ability to take the project from the detailed design stage through to completion. This includes experience of dealing with requests for information, managing design changes and controlling the costs associated with design changes. The role is usually advertised to include the following tasks:

- assess design details and design information for buildability;
- conduct value engineering exercises and implement innovative solutions;
- liaise with health and safety managers and environmental managers to ensure construction complies with previously determined requirements, such as town planning approvals, building regulations, CDM regulations and BREEAM;
- manage design changes and associated costs;
- manage information flow (provided on time) and requests for information;
- monitor design costs associated with the building process;
- review design information to allow construction to proceed safely and efficiently.

Relating this to the RIBA Plan of Work, the stages being managed at this stage include J, K and L for traditional contracts. However, depending upon the type of procurement route used, the construction design managers' role may have a wider span, including overlap with some of the earlier 'Pre-Construction' stages (F, G and H). In Case study 4C the role of a construction design manager provides some useful insights into some of the work that is undertaken at the construction stage.

Case study 4C

A design coordinator's perspective

Matt Griffiths – Thomas Vale Construction

Having graduated from Loughborough University's Architectural Engineering and Design Management programme in 2011, I went on to re-join the contractor that had provided me with a job as design coordinator during my industrial placement year. Thomas Vale Construction is a £250m contractor whose work is mainly, but not exclusively, based within the Midlands. Its 500+ employees are split across several divisions focusing on different sub-sectors of construction, with the office I work for (in Aston, Birmingham) covering mainly refurbishment, with some new build elements and retail, up to around £10m. Recently, Thomas Vale has been bought out by Bouygues Bâtiment International becoming part of its multi-billion pound group and a key factor in Bouyges expansion into the UK construction market. It continues to run under the Thomas Vale Construction name but now has the backing of the fourth large contractor in the world.

In my division, there was just one design manager whose responsibility lay in providing support to design and build tenders, while also supporting live jobs on the construction site. Having started six years ago, the design and build manager's role at Thomas Vale Aston became a stand-alone discipline. It is the supporting link between the consultants, site, buyers and estimators on jobs with extensive design elements. Since I started, I have worked alongside the D&B Manager to help resource the increasing number of design lead tenders and design and build jobs that have come in over the past few years. The design department has become an essential lynchpin in ensuring jobs are ready to be constructed efficiently and trouble free by the site team, often becoming the hub of information for all involved with the job.

Projects

With the size of jobs we work on it is not practical or necessary to be exclusive to one project at a time. Once on site a job will be run day-to-day by the site team with support and input where required from the design department. Some complex, intricate refurbishment jobs require more input than others by their very nature, but part of a design coordinator's role is to carefully manage and prioritise their time so to keep all their jobs running smoothly.

Currently I am working on five projects: two live jobs, one tender as part of a partnership with a council, and two competitive design and build tenders. I have also recently finished my first solo job which was a complete refurbishment of offices as part of a larger £2m package of a new factory's works. The live jobs represent two extremes so far as the contract values are concerned, with one being worth £700,000 and the other pushing £12,000,000. The larger is a project to transform Bath Spa railway station as one of the final stages of a major city centre shopping development by Multi Developments. It is a traditional contract; however, due to the size and nature of the works it has several 'contractor designed portions' involving heavy input from the design and build manager and myself. The smaller job is a fit-out of a newly built library in Shard End, Birmingham. Despite its lesser monetary value the intricate nature of the design works and the heavy coordination of trades mean that this tends to demand more of my time. It is the kind of relatively short-term project whereby a design coordinator is critical to meeting both client and timescale demands; and I have thoroughly enjoyed being part of nearly every package of works to see it come to fruition.

Roles and responsibilities

My job role is one of constant communication, information delivery and bringing people together to find optimal solutions to problems. For design and build contracts the design coordinator becomes the central link between everyone involved in the project. All the drawings, queries, problems and developments come through myself meaning we need to be up to date with nearly all aspects of the building. Probably the most important job I do is to cross check all incoming design information to ensure that the construction is fully coordinated and subsequently will be built without issues on site. Inevitably, these coordinational checks throw up all sorts of problems; from the incorrect setting out of partitions in relation to fixed furniture, to lighting clashing with smoke alarms. This is where my role is critical in bringing disciplines together to collaboratively resolve problems taking into account site restraints, designer's wishes and client needs. This is especially the case with refurbishment of existing buildings therefore often where two separate skill sets cross, the design coordinator is required to guide the resolution process.

The other side to my role is one of recording and tracking information. I am the link between the subcontractors and the consulting team, and therefore it is my responsibility to ensure that the work they are designing is approved by the consultants and all comments are taken on board before anything is manufactured or built. It is also a careful balance between ensuring designs are produced on time, but are produced correctly. In order to manage this process we record all the dates, statuses and details of all design information received and distributed via a combination of self-produced Excel tables and Thomas Vales own internal website which acts as a central database. This process is in place to prevent costly mistakes whereby the built works do not reflect what the design team and the client wanted or intended. It often involves extensive discussions or workshops with all parties in order to reach an optimal, workable outcome. Again we must bear in mind finances, site constraints and the clients' wishes to carefully craft a solution.

Education and training

Since joining Thomas Vale I have attended several in-house type training courses. The majority of my initial training was carried out 'on-the-job' during my placement year and I learned by example and by being given a slowly increasing level of responsibility. Since graduating, I have been given a training list with a selection of topics on which I have had to give a 'prioritised' rating. When a course becomes available, the people who ranked it the highest priority are invited to attend, and the places are allocated accordingly. The only other training days I have attended have been in conjunction with company policies (on things such as sustainability and PassivHaus) and some related to partnership processes (i.e. a new framework with the council). I am also continuing my CIOB chartership in the background, which is (at the moment) separate from company training.

What I like about my job

The best part of my job is seeing the drawings I have spent so many weeks working with slowly become a reality. It is satisfying to know that one has been involved in the production of a building where the space will be used by people every single day. Being a design coordinator means being involved right from the tender to the handover which is a really rare position to be compared to the rest of the Thomas Vale team who only see segments of projects. When I sit down with the architect, the end users and some sub-contractors to collaboratively resolve an issue, I really get a feeling of purpose in being able to help deliver a service to the people who ultimately will use it for decades to come. By acting proactively and diligently I cannot only ensure the project is delivered on time but in many cases improve the design through discussion, on the odd occasion even reducing the costs.

Room for improvement

There are some elements of the role which could be developed to enable design managers and design coordinators to switch focus, from sometimes being merely a 'post box of information' to spending time on really driving the job forward, spotting challenges and resolving them early in the process. A lot of my time is spent recording the receipt, delivery and status of information as well as reviewing the drawings against several other trades' drawings. This is a critical step which I described earlier, but essentially it could become an automatic process if all trades worked off a central 3D building model, as provided by Building Information Modelling. This would highlight clashes immediately and would prevent them from being discovered later in the process, often on the site, when there are considerable cost and time implications. It would provide design coordinators with a similar content of information to review and act upon, but it would eliminate the time consuming nature of manually inputting and checking that has to be done at present. The information extracted from the building information model by any party would be recorded automatically, which would render our manual records redundant, freeing up precious hours to focus on other areas such as managing sub-contractors designs and developing design information with the consultants.

Future directions

This leads on to what I believe to be the future of construction design management. 3D modelling is without doubt the next step in design coordination. It is already making headway in the industry with the biggest contractors, but with Thomas Vale targeting considerably smaller projects compared to the larger contracting organisations, the benefits are much less achievable at present. The jobs I work on are on the scale of a new school or hotel refurbishment, not Olympic parks or flagship skyscrapers. As a consequence, we use smaller subcontractors who sometimes have only just begun using email, let alone 3D modelling. Until 3D modelling is afford-able and useable for these types of businesses (SMEs), contractors similar in size to Thomas Vale will be unlikely to fully commit to BIM. I can see its prin-ciples being used to an extent such as virtual servers to centrally store drawings and creating a greater integration of many of the individual ICT systems used across the business, but the size of projects does not make full Building Information Modelling financially viable yet. However, increasing the collaboration and sharing of knowledge both internally with our own employees and to an extent with our subcontractors and consultants, would make my role in coordinating the design information quicker and much more efficient. As the industry evolves into an ever more electronically dependent one I can see the future of design management moving increas-ingly online and enabling design teams to work closer together via the sharing of a greater level of information than simply drawings alone.

Post-construction – maintaining design value

For both the new build and refurbishment market, there is a need to manage the built asset to maximise its service life. This takes us into the realms of main-tenance, asset, and facilities management, areas in which the link between good design and effective use of a built facility is well established. For the contracting organisations that offer design, build, maintain and operate type services it becomes even more important that design decisions provide long-term value to the building owners and users alike. Thus design management takes on a more central role in the organisation's business planning than it might compared to the provision of design and construction services.

Within the facilities management literature, there is a clear understanding of the importance of building design and its affect on the performance of employ-ees (building users) and the business. Similarly, the relationship between building maintenance and early design decisions is well established. As contrac-tors and consultants adopt a single virtual model (BIM) for their integrated project delivery there is the potential to link this information to the operation and maintenance of a facility. For repeat clients (such as Tesco) and clients with large built environment estates (such as universities) or large property portfo-lios (such as commercial landlords) the BIM is the source from which to make decisions about routine maintenance, alterations and future upgrades.

From the design manager's perspective this suggests greater communication and interaction with facilities managers and maintenance managers. This will allow precious knowledge about the client organisation and the performance of the property portfolio to be included in the briefing and early design stages.

Effective communication with building users will also be instrumental in helping to bring about a built asset that users are happy with. Indeed, there is a strong argument for bringing the facilities and maintenance managers and user representatives into the temporary project organisation to contribute to the co-creation of the brief and the design.

Conclusion

One of the most striking features of the case studies presented so far in the book is the importance of interacting with others to coordinate and manage design. Design managers need to have a thorough understanding of the entire project and be able to deal with the people issues. Being able to balance technical and task based issues with socio-emotional factors is an important skill; and one that has to be honed through experiential learning. In Chapter 5, emphasis is on the ways in which actors can get together to explore options, discuss and review design development. This naturally leads into Chapters 6, 7 and 8 where the emphasis is on information technologies.

Further reading

CIOB (2013) *Design Manager's Handbook*, CIOB/Wiley-Blackwell, Chichester.

5

DISCUSSING AND REVIEWING DESIGN

Construction design managers will spend a lot of time questioning the information provided by architects, engineers and specialist designers/suppliers/sub-contractors. This applies to projects that are managed using 2D information as well as those using a single virtual model; the difference being in how and when the information is reviewed and coordinated. And once discrepancies and errors have been found, it is important that the construction design manager can communicate on a number of levels and with a wide variety of participants. Pre-construction and construction design managers will be charged with coordinating information from many different sources, and regardless of the information communication technologies (ICTs) employed it will be necessary to interface with others to discuss the design and its realisation. This not only calls for effective processes and tools, but also the ability to communicate on and across many different levels and interfaces.

Communication media

Choice of communication media can play an important role in the ability of project participants to understand one another and hence work effectively. One medium may not necessarily be better than another, but it may be more effective at communicating a message for a given context and a given audience at a specific point in time. For example, plans and detailed drawings may be the design team's preferred choice of media, but the information codified within the drawings may be incomprehensible to clients who have no experience of construction. In such a situation, a physical or virtual model might be a better choice of media to explain key concepts and spatial arrangements. Similarly, individuals from different disciplines may have preferences for text and figures over drawings, or vice versa. Media may be used for one or more of the following purposes:

- *As an aid to the development of the design.* Media can be used as an aid to memory recall and decision making, so drawings, notes and diagrams are important tools for developing design ideas.
- *As an aid to coordination.* During the design phase, information is provided by a number of different providers, from manufacturers and specialist sub-contractors, structural and services engineers, to design, etc., to aid coordination.

- *For contract documentation.* Arguably the main focus of the production information, this is used by a variety of individuals to assemble the building.
- *As a design record.* Drawings and specifications will form the main part of the 'as built' documentation. Combined with maintenance information, operating instructions, warranties and guarantees this should be handed over to the building owner on completion.
- *As evidence in disputes.* Should a dispute arise during or after construction then the production information and any project documentation, e.g. letters and file notes, will be required as evidence, either to support or to defend a particular claim.
- *For facilities (asset) management.* As an aid to making decisions, such as space planning, maintenance, remodelling, etc., during the life of the building.
- *For recycling and disposal.* As a record document to aid with the effective and safe disassembly of a building that has exceeded its service life.

Communicators should consider a number of fundamental issues before choosing from the available media (summarised below). Of particular concern is the reason for the message: is it to exchange ideas or is it to convey instructions and/or information? The answer to this question will help in identifying the recipients of the message and the most appropriate, or effective, media to facilitate communications. Another issue relates to the formality, or otherwise, of the communication, and the choice of media best suited to convey and (if required) create a record of the communication. The choice of media also needs to consider the users and their ability to use the media in the workplace, be it an office environment or a construction site. Time is another determinant.

Oral communication

Oral communication skills are essential for explaining and defending ideas, exploring problems, coordinating work packages, resolving disagreements and getting things done. The ability to converse with others also has a social function in terms of developing effective working relationships. Discussions with our work colleagues form an essential part of our working day. They help us to establish information and help to provide the context for making informed decisions. Discussions are also used to help reach mutual understanding, build and reinforce relationships, test ideas, negotiate and reinforce mutual trust and respect. In a temporary project environment, face-to-face social interaction will usually occur in meetings, of various types and formality, and workshops, which by their very nature tend to be relatively informal.

Written communication

As part of day-to-day business there is a need to communicate in writing, thus the ability to write clearly and effectively for a variety of complementary purposes is an essential requirement. The level of formality may vary, for example between an email to a close work colleague and a formal instruction to change the design, but the purpose of the written communication is the same. The designer needs to be clear about distinction between information that is best conveyed in a drawing (or drawings) and that which needs to be included in a written document. Written communication includes emails, letters, reports and minutes of meetings, instructions and variation orders, request for information, specifications, schedules and contracts.

Graphical communication

Drawings are one of the most familiar and effective means of communicating information, although as with other forms of codified information the interpretation of the information can vary between users unless clear and accepted drawing protocols are followed. Standard conventions allow all users to quickly understand what is required. The main graphical tools used are sketches, 2D and 3D drawings, perspectives and virtual models which allow walkthroughs.

Physical representation

Physical models are an effective means of helping individuals to visualise spatial relationships and scale in three dimensions, for example via scale models and full size mock-ups. Physical models are used alongside virtual models to help develop designs and also to communicate with individuals who are less able to understand technical drawings, such as clients and members of the public. Sample panels are used on the construction site to check and approve samples of materials and the quality of work before the project commences.

Virtual representation

Powerful computers and computer software are now affordable for even the smallest design offices, providing the opportunity for networking, sharing, and coordinating information and the handling of vast quantities of information. The three-dimensional object based modelling systems provide a more user friendly design tool than the two-dimensional ones, which are essentially a drafting tool. Similarly, the development of BIM allows time (4D) and cost (5D) to be modelled within a virtual environment, allowing designers to collaborate, experiment and test ideas. The development of building information modelling (or alternatively building information management) software has allowed designers to be creative within a shared virtual environment. Being able to model how a building is assembled and subsequently disassembled can help to reduce uncertainty, resulting in safer and more efficient building processes. Greater certainty and clarity also has the effect of reducing the number of requests for information and also the number of requests for design changes.

Networking

Perhaps one of the greatest benefits of digital information is the ability to collaborate from remote locations. No longer is it necessary for design teams to share the same office space when they can be working on the same project from different geographical locations, linked via an extranet or intranet. The integrated service digital network (ISDN) comprises technological developments in such areas as fibre optics, satellite communications, broadcasting and digital transmission. Combined they form the electronic superhighways that offer instantaneous communication with high quality visual and audio resolution, ideally suited for the transmission of architectural images.

Meetings

Mintzberg (1973) found that managers spent nearly 70 per cent of their working day in meetings, the majority of which were scheduled and the rest impromptu. This finding tends to be supported in subsequent studies (Hartley, 1997). However, despite the amount of time and energy consumed by meetings they are relatively under researched within the management literature (Volkema and Niederman, 1995) and also within the construction literature (Emmitt and Gorse, 2007). Gorse (2002) found that there was a correlation between well-managed and chaired meetings and project success, with the most successful project managers better at chairing meetings and steering discussions compared to their less successful peers.

Function

In addition to facilitating the exchange of information, ideas and opinions and providing a forum for decision making, meetings are also used to:

- Appraise. Meetings are used to appraise progress and the performance of projects, organisations and individuals.
- Bond. Meetings fulfil a fundamental human need to communicate and bond, and hence help foster team relationships. They create a sense of belonging and reflect the collective and cultural values of the temporary project organisation. Meetings can also be used as a tool to help motivate the project team, although this function may be better served through facilitated workshops as discussed below.
- Control. Meetings allow managers to stay apprised of progress and in command of the tasks to be completed. They also allow those attending to follow-up information requests, allocate scarce resources, agree action and set deadlines. All decisions should be recorded in the minutes of the meeting.
- Coordinate. Face-to-face discussion may help with the coordination of works packages and the clarification of roles and responsibilities. The aim is to ensure that adequate resources are allocated to allow operations to take place effectively and safely.

- Develop trust. Addressing tasks and resolving problems in a meeting forum can help to develop trust between individuals as others are found to act with benevolence. Conversely, the failure of individuals to engage fully in problem solving will usually lead to (or confirm) a lack of trust. Either way, it is useful to know where the boundaries of trust lay.
- Explore possibilities and preferences, e.g. through structured client briefing exercises.
- Resolve and clarify. A timely meeting can help to resolve problems, differences of opinion, minor conflicts and disputes. It can also help to clarify certain aspects that might subsequently have led to unnecessary errors and reworking. This may be something as simple as misunderstanding how words are used or seeking clarification about apparent differences between the drawings and the written specification.

Many different types of meetings are convened during the life of an AEC project to serve a variety of complementary functions. These range from the informal to the formal, and the impromptu to the strategically planned. The main reason for holding a meeting is to bring people together in one place (physical or virtual) to discuss issues, and they can be used to:

- start projects;
- develop and maintain effective teams and groups;
- explore values and agree value parameters (e.g. briefing);
- discuss and review progress (e.g. of the design, the project);
- discuss and resolve disagreements;
- exchange information and knowledge;
- discuss and resolve problems;
- close projects;
- hand-over projects (or stages of projects);
- analyse projects (e.g. to gauge performance of the participants).

Participation and interaction

There are no guidelines on the number of individuals who should participate in a meeting, although as a general rule the larger the number of attendees the less the potential for all to participate fully in the discussions. Interactive media can be used in meetings to help reduce the amount of time spent describing issues and hence increasing the amount of time dedicated to discussing pertinent issues. Computer mediated workspaces, which include large interactive screens and laptops are being promoted as a means of improving communication and information coordination. Virtual prototyping provides the technology to discuss designs, constructability and the scheduling of construction work in large immersive laboratories. Access to these facilities is currently a barrier for the majority of projects since the facilities tend to be few in number and located at universities and research institutions.

Many decisions are made outside the meeting forum, either before it starts, in discussions during refreshment breaks, or after the closure of the meeting. These tend to be face-to-face discussions between two or three individuals anxious to reach consensus over a particular issue in order to present a united view and thus help avoid uncertainty, disagreement and conflict within the meeting.

Facilitated workshops

Workshops differ from meetings in that they are concerned with establishing and developing interpersonal relationships, either as a primary or secondary function of the workshop. Interaction is mainly socio-emotional. Development of relationships is often achieved by working collaboratively towards solving a (non-project specific) task (e.g. a simulated role-play exercise or an educational game) or by working collaboratively on a project specific issue, for example in a value management (or value engineering) workshop.

Function

In addition to helping to establish group membership and social identity in a temporary organisational setting, facilitated workshops are also used to:

- build trust;
- confront groupthink;
- create knowledge;
- develop working relationships;
- establish project parameters;
- explore different perspectives (and disagreements);
- resolve conflict.

Compatibility and values are difficult to establish from CVs and personal recommendations. It is not until people start to work together that the level of compatibility starts to become clear and values start to emerge through actions. Exploring the degree of compatibility through facilitated workshop exercises can be a highly effective way of accelerating individual understanding of their fellow contributors' values and preferences for communicating. Organisations may use 'awaydays' to allow their employees to understand one another better through non-work orientated activities in a workshop setting.

Workshops are a common feature of value management and value engineering exercises (see Case study 5A). Workshops provide a forum for systematically evaluating design proposals and exploring alternative solutions that may offer improved value to the client and project stakeholders. By bringing outsiders into the project it is possible to review the project objectively.

Facilitating workshops

Workshops, like meetings, need to be structured to achieve a specific aim. They should, where appropriate, be included in the project plan and their aims and objectives clearly stated. Unlike meetings, workshops tend to be used sparingly, often at the front end of projects to stimulate teamwork and develop trust, or at specific points in the project as part of a value management framework (see Kelly *et al.*, 2004 and Case study 5A). Facilitators have a significant role to play in creating an environment in which participants are happy to communicate openly and contribute to discussions candidly. To a certain extent, this is determined by the personality of the facilitator and his or her ability to bring about a trusting atmosphere very quickly.

Use of interactive media and gaming

Playing educational games is a well-rehearsed technique for developing interpersonal relationships. Interactive media provide additional opportunities to engage workshop participants in a collective task. Some facilitators will use toy bricks as a means of stimulating the participants to work together. Games based on solving problems and developing design solutions serve a similar function, to encourage socio-emotional development.

Collaborative planning workshops

One way of bringing people together is to develop the construction programme based on consensus of the participants. Collaborative planning workshops aim to do this by asking individuals to contribute to the planning of the programme. Usually conducted in one physical space (or alternatively synchronously via ICT from different locations) individuals are asked to commit to a specific task and specific timescale. They do this by placing a coloured label with their task identified onto a master programme (which is usually mounted on the wall). As others add their tasks and timescales there are inevitably some clashes and overlaps, which are discussed by those placing the coloured labels. The aim is to reach consensus about how the project is to be built, by whom and in what sequence. The facilitator's role is to encourage the participants to discuss problem areas and reach agreement. An additional role may be to encourage the participants to think about how they build and to consider different ways of achieving their objectives in less time.

This is a proven way of developing construction programmes that everyone signs up to, hence there is a sense of ownership of the whole process and hence a greater pride in making it work. An additional benefit is that the process often brings people together who might not otherwise have met face to face, and hence it may help the development of interpersonal communication and trust in one another. Case study 5A illustrates the role of a value management approach as a catalyst for enabling communication in projects.

Case study 5A

Value management: approaches for successful design management

Michael Graham, Managing Partner, UKValueManagement

Value management is one of those approaches which gets things done compared to just identifying what needs to be done. Over the years I have found this approach to be both versatile and extremely powerful. Deployed as a business framework, or within a project environment, value management directs and motivates people towards those activities which best serve the customer (client). Risk management enables the team to deal with hazards and uncertainty, but often something more is required to establish the goal. The value management approach does that and maintains performance towards the goal. As such, the approach can be very useful for design managers.

Benefits of value management

Value management is applied to motivate people, develop skills, and promote synergies and innovation with the aim of maximising overall performance. The approach can be seen as an investigative tool, design generator, conflict resolver, and commercial performer which can help to focus sustainable design and sharpen commercial performance. It:

- is a cross-cutting management system approach that directs and controls management activities related to value for the customer and the supply chain;
- integrates well with other management approaches and naturally fits within the team environment;
- helps people reconcile apparently conflicting objectives and competing interests to set an agreed strategy, establish team actions and make design decisions;
- uncovers and clarifies the functional requirements that must be satisfied by the design team;
- enables people to prioritise the relative importance of benefits for multiple scenarios (using value profiling);
- stimulates creative thinking to generate design options and enables effective decision making;
- delivers high commercial return on investment.[1]

There is always a strategic choice to be made. Should the design comply with the bare minimum requirements to gain regulatory approval at the lowest cost, do we do what we can with available resources, or do we do the best possible to enhance return for the customer? Some organisations position themselves as delivering the 'best', whereas others are associated with delivering commodities at low price. There are many balances to be struck between prioritising stakeholder needs and allocating resources to meet those needs. Value management gives the leadership team and the design manager in any organisation the framework, methods and tools to make effective decisions and deal with these complex problems and enable sustainable decision making.

Value and function

Two terms 'value' and 'function' are widely used (and misused) in everyday language. In value management, the terms are used to express fundamental concepts with specific meaning.

Value. Value is a measure of the beneficial return gained from the consumption of resources. In the UK, 'value' is often perceived when gaining a better price for a specified product or service. This is frequently more about 'cost reduction' than 'value improvement'. Value improvement arises when cost is reduced by eliminating wasteful or unnecessary function rather than by reducing the size of space or grade of finish. For example, sustainable design solutions which make use of natural ventilation and light and incorporate low carbon technology may take time and expertise to develop, but the long-term benefit throughout the whole life of the building can make the commercial case for the purchaser to invest.

Function. A function is an effect produced by activities or processes. It is most clearly expressed in terms of an active verb and noun, e.g. communicate information, develop skill, motivate team, establish requirement, generate idea, select option, implement solution, raise performance. A single design element may fulfil many functions: transmit load, delineate space, promote brand, and enable access. Function analysis examines the effects that the customer intends to create and the effects which the supply chain or the design must produce to satisfy those customer requirements. Unnecessary or detrimental functions (and the resource consumption associated with those) can be removed without any impact on customer service or product performance for the customer.

Function analysis clarifies the specific customer requirement which may be to control space, flex space, or flex operations. Frequently the customer is not certain of the design requirement so people tend to specify a larger space in the built environment. A functional performance specification such as '...On completion the design shall: flex space (within parameter ranges) and flex operations (within parameter ranges)' provokes much more creative thinking and much more information about what the customer requires than the product specification which calls for a large meeting room which can be split into two by installing a movable wall. The purchaser who specifies the product will be delivered that product at a reasonable price. Whether the design does the job intended or not is the purchaser's risk. The purchaser who specifies the required functions should be delivered a fit for purpose solution. Performance risk may still lie with the purchaser as the designer will use reasonable skill and care to create a design solution that

83

achieves the specified functions, but the purchaser who specified the performance requirements for building use will be much more certain of achieving a satisfactory design.

Value management includes three core methods focused on function:

- Function analysis (examining requirements and performance).
- Function cost (allocating cost to function rather than to building elements).
- Function performance specification (communicating the design requirement).

There are other methods which focus on value, including value engineering and design to (target) cost.

Practical applications

Successful organisations apply value management as a combination of management framework, methods, and effective use of tools.

Example 1: suppliers to a national house building company entered the annual round of framework contract price reviews, fearing a substantial price reduction target. Value profiling demonstrated that it is was much more important to protect sales. The annual review became an occasion to find ways to speed up delivery and increase options available for purchasers. As a result of substantial time savings, costs were significantly reduced without damage to margin per unit and a smaller than feared drop in sales was achieved in the market.

Example 2: a new food production plant was planned. There was significant disagreement within the business as to configuration of the main production lines. The decision had a dominant influence on the orientation of the building and the layout of the access routes. Products either flowed through the factory or arrived and departed from the same side. Value profiling highlighted the relative importance of low carbon emissions, flexibility for future expansion, and control of vehicle movements. Value management was established as a framework in the business and as a result the organisation's management had the option to call for a specific study of this problem. The project manager and design manager were working to very tight timescales and further delay in making the decision threatened the sales growth strategy.

In both examples the design manager worked with an independent professional in value management to plan a series of workshop discussions. These discussions involved all relevant stakeholders to respect value from the perspective of the ultimate customer and value from the perspective of the supply chain. By applying the 'standard' value management methods there was clear agreement on function: the effects which had to be produced. This gave the basis for useful examination of resource allocation (function cost analysis) and function performance specification for the minimum design standards and best in class design standards. Creative thinking then focused on how the performance specification could be delivered most effectively and decisions were then made. In both examples the impact of sustainable design solutions leading to low energy consumption through-out the lifecycle was recognised to be highly significant.

> Example 3: a prestigious public building was under construction. The contractor's design manager had difficulty gaining agreement between architect, structural engineer, and M&E contractor. Substantial modifications and breakout or new construction was being required at a late stage to accommodate plant and service access. Disputes were brewing and both physical progress and finan-cial progress were compromised. The design manager was focused on trying to apply the contract to impose solutions but the constraints were not in the control of any one party. This was a clas-sic case where partnering solutions were necessary, but absent. The main contractor's senior director had to make a radical change and called for a value management study, which was led by two inde-pendent professionals in value management. Function analysis and value profiling rapidly led to the conclusion that the site production schedule was the topic to tackle. Value management identified that, for this situation, clock time, not money, should be used as the resource measure. As a result of this intervention the whole supply chain was galvanised into synchronised action, working in harmony to deliver the project; recovering the poor physical and financial progress and eliminating future risk of delay (and costs) by arguing about who covers the cost of delay.

Conclusion

There are numerous international standards for practice that aim to combine various elements of good practice. The consistent theme between each international standard is the focus on function. Without a genuine understanding of the functional needs of the customer, no design solution can focus on what really drives value for the customer. Value flows from improving customer satisfaction (or benefit), effective use of resources, and productivity. Value management practice generates a sound understanding of functional needs which must be satisfied by design. Design is focused on meeting exactly what the customer requires: no more, no less.

The British and European Standard BS EN12973: value management sets out the approaches relevant for business in UK. I would strongly recommend that every design manager understands the implications of this standard, which builds upon preliminary work documented in European Commission Report, EUR 16096 EN, *Value Management Handbook*.[2]

Design critiques and reviews

The design review is a formal assessment of the design against the project brief. The aim is to check that the design proposal(s) meet the client's requirements, is in line with the design organisation's own standards, and conforms to the relevant regulatory requirements. Reviews usually take the form of a design critique or a more formal design review.

Design critiques

Design critiques tend to be relatively informal events conducted within the privacy of the (design) office. Designs can be discussed openly and critically with a view to improving the value of the design before drawings and associated information is released to the client and/or other project participants.

Design reviews

Design reviews are planned events, forming an important part of the programme and the project quality plan; essentially forming control gateways at pre-determined key stages of the project. Design reviews should include the presence of the client and consultants working on the project so that the project team reviews the design and any alterations agreed by the team and recorded in the office plan. These meetings provide an opportunity to discuss and agree the design before proceeding; more specifically they should address the following:

- design verification;
- design changes;
- statutory consents;
- constructability;
- health and safety;
- environmental impact;
- budget;
- programme.

The design review is a very good tool for detecting errors and omissions. It also provides a checkpoint for ensuring that the design meets the client's requirements and the architectural practice's quality standards. It also gives the planning supervisor an opportunity to check the scheme for compliance under CDM (Construction Design and Management) regulations. More importantly it provides a window for debate and feedback. It is important to keep these meetings organised, but as informal as possible so that ideas can be discussed freely and all members of the project can participate in the process. Planned design reviews where client, external consultants, designer, design manager, project manager and the planning supervisor can review, discuss potential problem areas, and take appropriate decisions, should form an essential part of a health and safety strategy. The design review has another purpose: the check for compliance with environmental/sustainable policies and practices. These may be a combination of the client's requirements and the firm's own pursuit of environmentally responsible policies and will have been discussed and agreed at the briefing stage. As the project proceeds, many situations arise and change, so therefore it is important to constantly review the project's environmental impact against the pre-determined criteria.

Design quality indicator

The design quality indicator (DQI) has been developed as a tool for assessing the design quality of buildings. The tool may be used at key stages in the development, realisation and use of the building and is well suited for use in conjunction with value management and risk management techniques. The four key stages are briefing, mid-design, at occupation and during use. The tool relies on participants completing a questionnaire that addresses the three fields of build quality, impact and functionality. The results are then expressed illustratively as a design quality indicator spider diagram. By using the tool at key points in the project, it is possible to track the importance given to all ten factors and helps to focus attention on areas that have not been adequately addressed. The tool is useful for developing and maintaining a clear vision for the building design. It also helps to capture knowledge for guiding future projects and the ongoing management of the building.

Requests for information

In an ideal world, the process of creating production information would be a smooth affair with everyone contributing their information on time, with the information received being complete, cogent, error free and sympathetic to other contributors' aims, objectives and constraints. In reality this is rarely the case, regardless of how good the managerial systems and the effectiveness of information coordination. Errors or omissions will result in a request for information (RFI).

Discrepancies between drawings, specifications and bills of quantities have the potential to undermine the smooth flow of the project as well as triggering disputes and conflict. There is a correlation between too much, or conversely too little, information and design changes. Well-planned and implemented design management process plans which incorporate regular design reviews can help to reduce problems related to the information. So too can the use of virtual models, where clashes and missing information can be identified before the design is approved for construction. Here, the argument for adopting BIM is very persuasive.

Design changes

During the course of a project, it is inevitable that changes will be required as the design evolves and more is known about what is required. Some changes may be relatively minor although others may have significant implications for the performance of the building. As a general rule, the further down the design process, the more significant is the design change. It follows that every effort should be made to establish the design before construction commences.

Changes to the design can come from a variety of sources: from the client, the design team and/or the contractor if involved early in the process. All changes need to be approved by the client before they are implemented, and organisations operate design management protocols that allow all requests for design changes to be recorded and considered before they are approved or rejected.

Changes during construction are wasteful of resources and in the majority of cases have significant cost implications. In new build projects, the only uncertainty should be the ground conditions, assuming that all of the information was complete and approved before work started. Once the ground works package is complete then there should be no reason for changes to the design or production methods unless the client requests a change. Design quality should be certain and cost forecasting should be accurate. In work to existing buildings there is more uncertainty since it is impossible to predict exactly what may be found when the building is opened up. The amount of work required and hence cost estimation is therefore less certain. Although every effort should be taken to limit the possibility of uncertainty at the production stage, it is likely that some changes may be deemed necessary. Changes to the agreed contract documentation will result in adjustment to the agreed contract sum. Most changes result in revising work and/or additional work as well as disruption to the programmed workflow. The inevitable result is an increase in costs, which someone has to pay for. Therefore it is necessary to track all requests for design changes and efforts should be taken to minimise the number of changes that occur during the realisation phase. Changes may be required for a variety of reasons; some of the most common are related to the following:

- Unforeseen circumstances. For example, problems in the ground or surprises when opening up an existing structure. This can be mitigated through extensive surveys prior to work commencing, but the risk of some unforeseen event cannot be completely eliminated. This is normally covered in the contingency sums.
- Client requests. These are normally related to clients revising their requirements, i.e. changing their minds, which can be mitigated by involving the client fully in the earlier design phases.
- Designer request. This tends to be related to the realisation that something could have been better and/or poorly conceived design work.
- Contractor request. Requests may be related to constructability issues and availability of materials to suit the programme. It is important to distinguish between those items that are a genuine problem and have to be revised

(e.g. clashes of services) and those requests made to suit the contractor (e.g. change of materials to save the contractor some money).

• Problems related to the information provided, resulting in requests for additional information and clarification.

Off-site production does not provide any opportunity to make changes once manufacturing has commenced, thus the design team and client must be absolutely certain that the design is correct before production starts. With site-based production there is always a possibility of making changes as the building is erected, assuming someone is willing to pay for the privilege. There may be considerable pressure to change the specified product and/or specified level of performance during the tendering and realisation phase of the contract. Most changes are formally requested and approved before being implemented and subsequently recorded in the as-built documentation. However, there is evidence that unscrupulous contractors and sub-contractors may change specified materials and components for cheaper alternatives and not inform anyone. A vigilant clerk of works can help to prevent some of these unwelcome habits, so to can the employment of reputable contractors and sub-contractors.

Design change management

Changes, regardless of their origin, need to be referred back to the design manager and checked against critical documents such as planning approvals and the project brief before implementing them. Requests to change building products and details have implications for the durability of the building and must be given careful consideration before a decision is made. In many cases this is not a quick process, since the changes may have implications for other interconnected aspects of the building. This means that the manager of the construction contract must make requests for changes in adequate time and be prepared to wait for an informed decision. Contracts stipulate clear rules and timescales for requesting and responding to changes. All approved changes must be recorded and the drawings, specifications and schedules revised. This will ensure the 'as-built' drawings are an accurate record of the completed building. Changes made without the knowledge of the contract administrator will not of course be recorded in the as-built information.

Managing interfaces

Bringing disciplines together helps to ensure that the various aspects of a project are developed in a manner befitting the client's aspirations. However, creating a TPO also creates a vast number of interfaces between organisations, organisational departments, teams, groups and individuals. Some of these interfaces will be new; some may be re-established after a (short) period of not working together and some may be continued from previous projects. Effort expended in forming relationships, testing new acquaintances' abilities and trying to establish their trustworthiness at a very early stage in the project can be beneficial in quickly establishing effective working patterns. Trying to establish compatibility as early as possible in the life of a project can be a time consuming task, but this effort can have significant dividends in terms of quickly building effective working relationships, developing mutual trust and reducing the likelihood of misunderstandings and unnecessary conflict (Emmitt, 2010).

Interface management

Interface management is concerned with the relationship between inter-dependent organisations working towards a common goal (Wren, 1967). This includes the identification of interfaces, the ability to clearly define responsi-bilities and also ensure effective communication across the interfaces. Given that the number of interfaces can be significant with a TPO it may be useful to consider two fundamental types of interfaces that are present within projects:

- Organisational (business) interfaces
  Organisational interfaces are mainly defined by contracts and the project context. Inter-organisational relationships are concerned with organisa-tional culture and the interoperability of management and ICT systems. Although the relationships can be dynamic, they are relatively straightfor-ward to define, map and manage through the life of the project.
- Personal interfaces
  Individuals interface with others representing other organisations, not the organisation *per se*, thus interfaces are coloured by the ability to commu-nicate and work with representatives of other organisations. Effectiveness of the relationships is dependent on compatibility of the individuals concerned. These interfaces are more challenging to define, map and manage, because over the course of a project it is not uncommon for indi-viduals to move jobs or be allocated to different projects, thus introducing new individuals and interfaces.

An example of how an interface management approach helped a small contrac-tor firm to effectively define and manage project goals is discussed in Case study 5B.

Case study 5B

Addressing the designer–contractor interface

Stephen Emmitt

A small design and build contractor formed the focus of research into inter-faces between contractors and designers. The company employs its own workforce of project managers and site operatives, which has allowed the company to maintain and deliver work to high quality standards, through which it secures repeat commissions. Based on the company's reputation for quality, the majority of its work comes through cost negotiations with clients, rather than through competitive tendering. However, it is common for the contractor to be appointed sometime after the architects and structural engineers (usually after town planning consent has been granted). The effect of this is that the contracting organisation has little or no choice over the design consultants; hence it is usual for the organisation to work with new, unfamiliar, consultants on each project. This means that the organisational and personal interfaces are different for each project. The challenge for the contracting organisation is to manage these interfaces effectively to ensure an effective working relationship and the effective flow of accurate design information.

The architects' perspective

Interviews were held with architectural practices to try and establish how the working relationship between the architects and contractor could be improved. Although the architects reported good working relationships with contractors, they raised a number of similar themes that fell under the theme of communication and design management.

Communication

The architects expressed the desire for better communication between the contractor and their offices, claiming that the way in which the projects were procured, appointing the designers long before the contractor, were not conducive to effective communication and collaboration. Their desire was for one or two meetings early in the project to discuss the scheme, with the aim of exploring and resolving uncertainties with the design. The architects claimed that the failure of contractors to make decisions early in the process often caused problems with completing their production drawings, often resulting in unnecessary work to accommodate late decisions. Examples given included:

- Not placing the order for the lifts (elevators) early in the contract period. Until the lift order has been placed with the lift manufacturer, the architects cannot get definite information on lift-shaft sizes which prevents the completion of the production drawings. They felt that contractors delayed the choice of lift manufacturer, and hence the placing of the order, until too late in the contract.
- Not deciding whether imperial or metric door sets were to be used. Because these door sets differ in size, it is not possible to complete the production drawings without a decision.
- Not deciding on the kitchen layout, which delays the completion of production drawings and also delays the positioning of water, gas and electricity points.

All three issues could be regarded as rather minor in the overall context of the project, but combined the uncertainty around these three areas prevent the architects from completing their drawings and hence delay the issue of information to the contractor.

Practical measures

These findings led to a number of practical interventions being implemented by the contracting organisation. A design management protocol was designed and piloted on a number of projects before being applied to all projects. The protocol set out a number of procedures to be followed by staff when dealing with external designers. The aim was to improve communication and introduce opportunities to stimulate collaboration between the contracting organisation and the design team. Two initiatives were instrumental in helping to achieve better collaboration:

Orientation meetings

One way of improving communication was to arrange an 'orientation' meeting at the start of all projects. This included personnel from the contracting organisation, the architects and the engineers. An important requirement was that the individuals who were going to be producing the information attended the meetings (not just their line managers). The aim of the orientation meetings was: first, to identify areas of concern about the production drawings, second, to discuss the contractor's approach to design management and hence familiarise the consultants with their culture. Feedback from the design consultants and the contractor indicates that this was a successful management intervention.

Collaborative planning workshops

Collaborative planning workshops were introduced to address the flow of work. These are held prior to the commencement of the project, and soon after the initial orientation meeting(s). The aim was to plan the construction programmes collaboratively, drawing on the expertise of those contributing to the workshop. By discussing the flow of work and interactions between trades, the intention was to try and save time and hence condense the programme. A secondary aim was to bring the main participants together to meet, share their knowledge and become better acquainted with the lean thinking philosophy. By engaging in the workshops, the participants had the opportunity to develop working relationships with the other project contributors.

One of the directly observed benefits of the workshop was the interaction between the designers and the trade foremen (who normally rarely met). At the start of the workshop, the body language of the participants was defensive. However, as participants started to relax, their body language became much less defensive and they started to discuss the project quite openly. It was evident that the collaborative planning workshops started to break down personal barriers, resulting in less defensive behaviour and interpersonal communication between the architect and trades foremen for the first time. This had a direct benefit of helping to improve the clarity of some of the drawings provided by the architects and led to new (innovative) ways of addressing the construction work.

Discussion

The organisations involved are small and all have limited resources on which to draw: hence the need for simple and effective design management procedures. Tools such as collaborative planning appear to be well suited to small organisations because they not only help the project participants to understand the scope of the project, but they also help to emphasise the requirements of the client to the project contributors. This helps to make client value more visible and the removal of inefficient processes more effective. Participants have also developed a better understanding of what is required and the ability to identify the need for change and process improvements. This is critical as an important contribution to the longer term sustainability of the business.

Conclusion

The issues addressed in this chapter are often referred to as 'soft issues' because they are concerned with people and how they interact. As highlighted in the earlier case studies, design managers will spend a considerable amount of time dealing with people and interfaces, primarily by managing and attending some form of meeting or workshop. They will also spend a great deal of time reviewing designs for compliance, managing requests for information and addressing the impact of design changes. With the uptake of ICT and BIM, the role of the task of the design manager is, on the surface at least, becoming a little easier. In the chapters that follow, our attention turns to collaboration technologies and BIM, before returning to strategies for collaborative working (Chapter 8).

Notes

1   Returns on investment of order 20–40 times direct cost and higher are reported (www.valueforeurope.com).
2   At the time of publication the *Value Management Handbook* was available as a free PDF download at http://bookshop.europa.eu/en/value-management-pbCDNA16096.

6

COLLABORATION TECHNOLOGIES

Construction project teams use a range of collaboration technologies to share and manage project information through the lifecycle of a project. Examples of collaboration technologies are intranets, extranets, podcasting and vodcasting technologies, BIM (Building Information Modelling) tools and tools for virtual-reality and video conferencing, among others. The purpose of this chapter is to introduce the different collaboration tools that are used to manage construction projects and consider how these tools can benefit the entire construction process.

Construction collaboration technologies

Construction is a complex industry involving a wide array of disciplines participating at different stages of the construction process. A standard construction project involves several disciplines working together towards a common goal – a completed construction project. It is a team effort, which involves several, inter-organisational activities, dialogues and information exchanges. These teams involving the client, architect, structural engineer, civil engineer, contractors and subcontractors, perform diverse functions and activities during a project's lifecycle. It is this multi-faceted nature of the construction industry that makes the construction processes difficult to manage.

The nature of the construction sector is such that fresh teams are formed for almost every project, and these teams usually disperse once the project is completed. Each team is unique with respect to the number of members, type of disciplines and size of the organisation. The ICT (Information and Communication Technology) tools deployed during the project lifecycle may vary from team to team as well as from project to project. Each organisation has varying levels of ICT uptake and the software used; further to which the software applications used may vary from one construction project to another. Organisations and individuals participating in the team will often bring their own unique skills and resources, which may include legacy data and applications. The entire process is an amalgamation of different disciplines and activities. All these factors collectively contribute towards the complexity and fragmentation of the construction sector. The complex and fragmented nature of construction has been highlighted in several government (Egan, 1998 and Latham, 1994) and research publications (Lottaz *et al.*, 2000, Obonyo, 2001,

Ruikar *et al.*, 2001). The Egan Report (1998) stated that fragmentation of the construction sector is caused by the wide-range of disciplines involved during the construction process. Regardless of these factors, construction projects share some common features. Construction projects are one-off, unique, finite, goal-oriented ventures that are undertaken in real time (Allinson, 1997).

Although the construction process may appear to be linear in structure, it is in fact a combination of iterative and linear construction processes. It is therefore very important to plan and manage the entire process efficiently and effectively. Effective management of construction projects requires vision, ambition, some forms of inception, planning, monitoring, control, coordination and synchronisation of proposals, allocation of roles and responsibilities and some final closure that brings the project to an end. If construction projects are to be successful, reasonably economic and hassle free undertakings, they must be managed purposefully (Allinson, 1997), and should be aimed at providing a value added finished product to the client. The delivery of a 'value added' construction project that meets or even exceeds the client's expectations requires effective and efficient management of the entire process, from start to completion. With growing complexity and the ever present pressure to reduce the duration of the construction programme, the need for efficient collaboration is becoming more and more important. Such collaboration may be hindered by the lack of effective and efficient facilities for exchanging and organising project information (Lottaz *et al.*, 2000). However, technology and science are growing at a rapid pace and driving the construction sector to explore and develop new and innovative tools to overcome the inefficiencies and hence improve coordination and collaboration.

At the grassroots level, collaboration technologies are ICT based systems that enable collaboration between two or more individuals, or groups of individuals, at inter- and/or intra-organisational levels. Wilkinson (2005) defines collaboration technologies as

> a combination of technologies that together create a single shared interface between multiple interested individuals (people) enabling them to participate in creative processes in which they can openly share their collective skills, expertise, understanding and knowledge (information), and thereby jointly deliver the best solution that meets their common goals, while simultaneously creating an auditable electronic record of the people, processes and information employed in the delivery of the solution(s).

Intranets and extranets

Intranets

An intranet is a secure computer network that spans an organisation and connects its people and information systems across functional and geographical boundaries using Internet Protocol (IP) technology. Some intranets can only be accessed from a computer within the local network; however, some organisations provide access to remote employees by either using a virtual private network (VPN), or using access methods that require user authentication and encryption.

Access to an organisation's intranet is normally restricted to its employees only. Typically, an organisation's intranet holds organisation specific information related to standards, policies, projects, processes and people. For example, using an intranet company staff can seek expert staff listed on their internal skills database. Some intranets are also used to deliver specialist tools and software applications to aid specialist activities. Examples include corporate directories and tools for business travel management, supply chain relationship management tools, enterprise resource management and project management. Some organisations use intranets as knowledge vehicles, where existing knowledge is either stored (in databases) or new knowledge is generated (via topical discussion forums or online communities of practice).

Extranets

An extranet is frequently defined as a variant of the intranet (Wilkinson, 2005). For example, an extranet can be seen as an organisation's intranet that extends access to external users (e.g. project partners, suppliers and vendors) outside the organisation, in a secure and controlled environment. In a construction project context, when the controlled access extends to project teams members, the extranet is often referred to as a 'project extranet'. Thus, a project extranet provides an interactive and secure environment in which different organisations involved in a construction project can exchange project information, documents and other project related data in a seamless and secure, Internet-enabled environment. There are several challenges that should be considered before using interactive information sharing systems such as project extranets (Liston *et al.*, 2001). The project extranets should:

- Be useful over an array of project disciples and integrate with project member's legacy systems;
- Make use of legacy sources of information that were not initially designed for multi-disciplinary settings;
- Provide a uniform representation that has enough semantics to provide useful structure while being flexible enough to be usable by multiple parties who bring different perspectives and representations to a project team; and
- Integrate structured information (of the kind in model databases) with unstructured information (text).

Project extranets facilitate auditable management of construction projects. Using project extranets, project teams can easily access project information from anywhere at any time in real time. Some of benefits of project extranets are faster transaction times, better transparency in the exchange of project information, improved collaboration between construction project partners, time savings for communication of project information, savings on project cost, and streamlined construction business processes (Ruikar *et al.*, 2001). There are several extranet tools that are specifically designed to meet the unique needs of a construction project. Table 6.1 lists examples of the construction-specific project extranet tools.

Table 6.1 Examples of project extranets tools

Company	Web address/description
Asite	http://www.asite.com/ Software as a Service (SaaS) extranet environment with hosted applications and solutions specifically designed for the construction industry.
Conject	http://www.conject.com A SaaS provider offering secure web-based collaboration and project management system that enables professionals to communicate and manage information about built assets securely over the internet.
Bricsnet	http://www.bricsnet.com/ An online project collaboration tool throughout the lifecycle of a building.
Autodesk buzzsaw	http://usa.autodesk.com/buzzsaw/ Online collaboration, printing and procurement applications for global design, construction and property management industries. Allows members to view files and order prints online, providing seamless integration from the design desktop to the digital print room.
Cadweb	http://www.cadweb.co.uk Cadweb enables through-lifecycle sharing, retrieval and management of large volumes of construction project information.
Causeway	http://www.causeway.com/ Causeway is a global SaaS provider that offers a suite of software tools that support the complete lifecycle of the built environment.
Sarcophagus	http://www.sarcophagus.co.uk/ Sarcophagus is an online project management tool that is used to aid business project collaboration. Other services include tools for email management and online tendering.
TendersDirect	http://www.tendersdirect.co.uk/ TendersDirect is a tender alert system that provides access to current government and utility company contracts, both within UK and in Europe including a searchable database of tender documents.
The Building Centre	http://www.buildingcentre.co.uk/ Single-source information repository covering aspects of architecture and design, construction and planning, home improvement, DIY, and self-build.
4Projects	http://www.4projects.com/ An on-demand SaaS extranet environment for collaborating with disparate stakeholders.

Podcasting and vodcasting

The term podcasting evolved from the words *iPod* and broad*casting*. Podcasting is the process of capturing an audio event, editing it through a program on a personal computer, and then posting that digital format to a website in a data feed called an RSS 2.0 (Real Simple Syndication) envelope for publishing content that is frequently updated (Meng, 2005). To access a podcast, users can subscribe to the website which contains the RSS 2.0 feed through an audio management software (such as iTunes or Windows Media Player), which can automatically download the file onto their computer. Additionally, users can synchronise their portable audio device/s with the audio management software on their PC to transfer the file to the portable device and access it as per convenience.

More recently, another term vodcasting has emerged, where *vod* is an abbreviation for video on demand. Vodcasting is a step forward from podcasting. It is similar to podcasting with one key distinction that separates the two, in that vodcasts support both, digital video and audio formats (Herkenhoff, 2006). This difference changes the specifications of the portable device from a portable audio device to a portable video device, which has a screen to view the audio video recording. However, as the technology develops and newer functionality is added to portable media players, it is possible to incorporate audio visual content into a podcast. For example, an iPod Touch has functions such as external volume controls, built-in speakers, Bluetooth support and presentation software. Its new features are faster hardware (microprocessors, graphics engine and RAM), voice control and a microphone. Thus, such podcasting devices have the potential to broadcast dynamic, audio visual content such as music files, voice recordings, presentations and photos. Podcasting combines elements of several disparate technologies: audio visual recording and editing, content syndication, and internet file transfers – into a single seamless process that retrieves audio from a website onto a listener's computer and synchronises it with an external digital audio player (Affleck, 2009). Table 6.2 provides an overview of the various tools needed to create audio visual podcasts.

Both podcasts and vodcasts have potential for widespread application in construction. A case study of how podcasting was used as an effective knowledge capture and sharing tool in a project based environment helps to demonstrate the value of the technology.

Table 6.2 Podcasting tools (Meng, 2005; Herkenhoff, 2006; and Kajewski, 2007)

Required tools	Description
Internet	Internet is required to publish the content, subscribe to the RSS feed and to access the content. Due to the large file sizes and the amount of information needed to be transmitted a broadband internet connection is essential.
Website	To secure the information and publish podcasts to project-stakeholders, a website with an RSS feed is desirable. Access to podcasts could be controlled by assigning usernames and passwords.
Computer	Standard laptops or PCs with internet access. This project used an Apple Mac with standard processing software such as iWork. Synchronisation can be done on a PC, a laptop or a portable device such as iPod Touch.
Audio video recordings	The tools required to capture audio visual content will depend on the purpose of the podcast, which affects the quality of the content. Audio recordings require input devices such as a microphone. Audio production software such as GarageBand is suitable for use in a Mac OS. Remote interview and conference recording, is possible with GarageBand. GarageBand provides flexibility in mixing different tracks and modifying the properties of each track (including filters, equalizer settings and echo). The program processes these changes simultaneously as it plays back the audio, without altering the underlying audio data. Permanent edits can be made to remove gaps, speech stumbles and other unwanted content, as well as cut, copy and paste audio bits into other locations.
Graphics and presentation	A variety of software can be used to create and edit graphics. For example, iWork Keynote is suitable for creating high-quality presentations with animations. Animations are desirable as they draw attention to the points of discussion and engage learners.
Synchronising audio visual content	Using GarageBand audio content can be recorded and then synchronised with the visual content in Keynote using a feature which allows users to drag an audio file onto a slide and then playback the recording when viewing slides.
File transfer software	The content files will need to be published on the web using file transfer methods including FTP/SFTP, HTTP upload.
RSS Enclosures	To prepare the content for delivery to the website it has to be tagged via XML in an RSS 2.0 format. A software designed to create RSS feeds can be used.
Portable output device with suitable hardware	A laptop and other portable digital media players can be used to access podcasts. For podcasts with high quality graphical content, a PC or a laptop will be most appropriate. This is primarily due to the widescreen and the functional playback environment.

Case study 6A

Podcasting

Kirti Ruikar

Case study background

This case study explains how audio visual podcasts were used to capture and utilise architectural design knowledge of a master planning project in the UK. This was done to retain the 'project memory' by capturing lessons learned at the early design stages. The podcast content included commentary from the lead architect of the project, who explained the design decision rationale that led to the development of the concept. Details such as the functional requirements, spatial organisation, the response to design brief, design decisions and drawings, and other related information, were included for context. The visual content included conceptual design drawings, site photographs and other relevant graphics that were necessary to visually depict the outcome of the design rationale to teams joining the project at a later stage. This had the potential to alleviate the problems of knowledge loss at downstream construction phases by enhancing the team's understanding of the issues and considerations that led to the final design.

Why capture knowledge in case study project?

Given that phase one, i.e. the conceptual design of the master plan, had only recently been completed, the design team was still intact and records of design data were easily accessible. This is important because quite often complex master planning projects, such as the case study one, last several years and involve a large number of stakeholders who form transient design teams through the project lifecycle. When teams disperse there is a risk that vital design knowledge will be lost. To counter this risk an archive of the key design decisions that influenced the conceptual development is clearly advantageous; especially since subsequent design teams are likely to be better informed about the design rationale and therefore take measures to ensure that the core project ethos defined at the conceptual stage is not lost in transition from one design team to another as the project develops.

How to create an audio visual podcast?

To create an audio visual podcast of the design project, a number of stages are involved. What follows is a detailed account of the stages involved, starting with sourcing podcast content through to publishing and accessing podcasts (see Figure 6.1):

Figure 6.1 Stages involved in creating audio visual podcasts of projects

1. *Identify sources of knowledge leaks and losses*: this is an important stage, which involves dialogues and discussions with key project stake-holders to identify the main sources of knowledge leaks and losses. The purpose of these early discussions is to identify ongoing or completed projects and capture the design team's perspective (i.e. decision ration-ale) about the development of the project. The design decision rationale that is captured is of value because knowledge is not lost, but retained for future reference.

2. *Obtain audio visual data for podcast*: this stage involves identifying a project, or projects, that would make up the audio visual content of the podcast. Typically this would include project specific data such as details of functional requirements, spatial organisation, the design brief, design decisions, design drawings, site photographs and other related information that would convey the ethos of the project and its devel-opment. The sources of audio content (commentators) would vary in numbers depending on the nature of the problem and the involvement of various team members in taking the decisions.

3. *Create audio visual podcast*: this stage involves creating an audio visual podcast. In the case study the audio visual podcast content was created using Apple's presentation software, Keynote. GarageBand, an audio production software, was used to record the lead commentator's discussion. Using GarageBand, an audio segment of each slide of the presentation was recorded and the content saved as an audio file. This was then synchronised with the presentation in Keynote using a feature that allows users to drag an audio file on to a slide and then play back the recording when viewing the slides. The presentation was improved by adding animations to draw attention to key discussion points. For the audio visual content to be podcast, it is necessary that it is playable in a portable audio visual device such as an iPod Touch. The Export option in Keynote enables this process, by allowing the audio visual presentation to be saved as a podcast.
4. *Publish podcast*: at this stage the podcast is published (broadcast) on the project server so that registered project teams may access its content.
5. *Download and 'learn as you go'*: the download function allows the podcast file to be saved in the iTunes library and to synchronise iPod Touch (or other equivalents) with the computer. iTunes helps to add digital audio visual files from a PC or Mac directly to the iTunes library. Video content can be added in QuickTime or MPEG-4 format to iTunes.

Was it worthwhile?

Yes, because the knowledge rich audio visual podcast enhances the learning experience of the project teams by capturing the underlying project ethos. Key project knowledge is retained, not lost. The audio visual podcasts is also reusable as a knowledge repository of considerable value to the learning experience, given that experts' knowledge is accessible on demand, offering flexibility to access knowledge independent of the experts' availability. The portable content of the podcast is downloadable on demand and provides context through the clear narrative of 'design history'. The commentary coupled with the dynamic graphical podcast content (i.e. design drawings, maps and photographs) creates a rich visual picture of the project.

Building information modelling (BIM)

BIM is a digital representation of physical and functional characteristics of a facility (BuildingSMART, 2012). It enables complex building and project information to be modelled into a single, virtual, model that represents all elements of the building or project. Thus BIM is about building (virtually) before anything is built (physically). BIM could, therefore, be described as a visual-knowledge base built into a digital n-dimensional intelligent CAD model.

Building information models contain accurate and precise geometry with relevant data (properties) to aid the design, construction process and the life-cycle management of a facility. Other dimensions such as time (4D), cost (5D) and nD supporting energy analysis of buildings and facilities management functionalities have been developed as additional bolt-ons to improve the true representation and accuracy of the simulation. BIM provides an 'intelligent picture' of a facility that not only holds information of the physical attributes of the building, but also the intelligence associated with it (e.g. decision data). The physical attributes relate to the building and its components and include details such as the physical attributes of components, their functions and chemical compositions. Intelligence refers to knowledge about the building and its individual components. Typically it includes rich data related to why a component was selected, by whom, lists of suppliers (where it can be sourced from and why the suppliers were selected), how it relates to the project's programme, and so on. BIM is a shared repository of knowledge about a facility forming a reliable basis for decisions through its lifecycle from inception to post-occupancy and onwards to the future disassembly and materials recovery phase.

During the last decade there have been major developments in BIM and its application to the design and construction processes. For example, BIM models can assist by using 3D for space planning, which has been used successfully for interior space management. Events such as office relocations can be designed before implementation on site. The cost of each space can also be calculated from the model via an extracted schedule (Boyce, 2009). BIM models can also be linked to facilities management databases and software to enable highly efficient and effective aftercare and maintenance of the building. Model-based FM software, Maximo and COBie are initiatives VINCI Construction UK Limited have implemented (Pang, 2010). One of the key features that BIM models can contribute to the design management task is by ensuring that all design inputs are coordinated with the use of clash detection. The clash detection capability, which exists in reviewing software such as Autodesk Navisworks, has reduced the number of design clashes that are 'discovered' during work on the site (Khemlani, 2012). This greatly assists design managers in their coordination tasks and helps to reduce the number of requests for information, design changes and delays on the construction site, leading to more efficient, cost effective, and safer working. Clash detection software enables the design team to review the entire model by examining all instances of structure components and the interface with non-structural elements to ensure connections and joints have been designed correctly. BIM is also used for fabrications, with software such as XSTEEL (Tekla, 2012). Steel fabrication is a prime example where the construction sector can design in 3D with either standard size structural members or bespoke design, which can be extracted and sent directly to the workshop for fabrication.

Some of the best known and used BIM tools that are currently used in the UK construction industry are MicroStation Triforma (Bentley), Revit (Autodesk) and ArchiCAD (Graphisoft). A detailed account of how these BIM tools are used in practice and what impact BIM has on design management is included in Chapter 7.

Virtual reality tools

In the context of the built environment, Virtual Reality (VR) involves the three dimensional generation of a computer model that represents the complete built environment. Such a VR model not only visually depicts the building environment, but is also interactive allowing users to 'virtually' experience (i.e. walk through) the building from various angles. This visualisation capability is useful because it allows clients to understand the design better than they would have if presented with static 2D design drawings. VR is used within construction for various design applications, for collaborative visualisation and as a tool to improve construction processes (Whyte *et al.*, 2000). According to Whyte *et al.* (2000) VR forms a natural medium for building design as it provides 3D visualisation, can be manipulated in real time and can be used collaboratively to explore different stages of the construction process. VR plays a pivotal role at different stages of a construction project. For example, because of the ability to thoroughly examine the design before construction work begins on the site, the design teams have the option to improve the design by identifying clashes and inconsistencies early on. It is a powerful, visual tool for showcasing concepts and ideas to potential clients.

For example, VR has been used to model the flow of people (e.g. crowd movement patterns) in order to define building design parameters (corridor widths, dimensions of stair risers and treads, position and width of doors, etc.) and device effective evacuation strategies in case of emergencies such as fires and terrorist attacks to ensure public safety (Varughese *et al.*, 2010). Another example is provided by Lipman and Reed (2003) who demonstrated how a large multi-national engineering company used the VRML models, generated from their CIS/2 (CIMsteel Integration Standards Release 2) files, to coordinate the work processes of designers, detailers, fabricators, and erectors based in three countries for a large process plant. Each of the project participants was able to view the same VRML model without the need to invest in proprietary software. The VRML model also contained links, in the text popup window, to traditional CAD drawings.

VR has different modes of displays. These include Head-Mounted Displays (HMD), Binocular Omni-Oriented Monitor (BOOM) displays and CAVE. HMD is a display device, worn on the head or as part of a helmet that has a small display optic in front of one (monocular HMD) or each eye (binocular HMD). A typical HMD houses two miniature display screens and an optical system that channels the images from the screens to the eyes, thereby, presenting a stereo view of a virtual world. A motion tracker continuously measures the position and orientation of the user's head and allows the image generating computer to adjust the scene representation to the current view (Beier, 2008). As a result, the viewer can look around and walk through the surrounding virtual environment. To overcome the uncomfortable intrusiveness of a head-mounted display, alternative displays such as BOOM and CAVE for immersive viewing of virtual environments were developed (Beier, 2008). The BOOM is a head-coupled stereoscopic display device. Screens and optical systems are encased in a box that is attached to a multi-link arm. The user looks into the box through two holes, sees the virtual world, and can guide the box to any position within the operational volume of the device. Head tracking is accomplished via sensors in the links of the arm that holds the box. CAVE was first introduced in 1992 by the University of Illinois. A typical CAVE is a room constructed of large screens on which the graphics are projected onto two to three walls and/or the floor. This creates a high resolution, multi-person, room sized, 3D audio visual, immersive environment. In a typical CAVE, graphics (landscapes, building interiors, etc.) are rear projected in stereo onto two walls and the floor, and viewed with stereo glasses. When the viewer wearing a location sensor moves within the CAVE's display boundaries, the correct perspective and stereo projections of the environment are updated, and the image moves with and surrounds the viewer; thus allowing a truly immersive experience. All of these technologies can greatly assist the design team with the development of complex designs.

VR also supports collaboration. For example, two or more networked users at different geographic locations can virtually meet by using one or more modes of display (i.e. a HMD device, a BOOM device, a CAVE system). Each user has a virtual identity (avatar) and each user views the same virtual environment from their points of view, depending on the display device they use. In this virtual shared environment the users can see each other, communicate with each other, and interact with (or within) the virtual world collaboratively as a team. Due to its capability to visualise and experience the real environment virtually, VR has several applications. Some examples from construction include the use of VR to train for safe assembly and disassembly and operating of site equipment, evaluating designs (for fire safety, structural integrity), architectural walk-throughs (visualisation and clash detection), human behaviour in built environment (e.g. impact of light and colour on productivity), simulation of assembly sequences and maintenance tasks.

Video conferencing

Video conferencing enables two or more distributed participants to communicate and share audio, video and textual data over a telecommunication network. Although face-to-face meetings are an important aspect of developing business relationships, in recent years virtual alternatives such as video conferencing have gained popularity. This is because video conferencing overcomes some of the difficulties associated with face-to-face meetings. For example, the inconvenience of international travel for face-to-face meetings, such as long waits at airports, lack of space in planes, lack of flexibility of flight schedules, flight delays or cancellations, limited network connectivity for cellular phones abroad and the very high costs of using phone services provided on plans, have together contributed to the growing popularity of video conferencing tools. Besides, a video conference is seen as a viable, cost effective alternative to business travel and is an important aspect of modern-day communication and collaboration. Factors such as globalisation, advances in information and communication technologies, high capacity broadband telecommunication, speed of the internet, easy access to powerful computing processors, ease of use of video conferencing tools, cost savings (due to the substitution of certain business trips) and their 'wider' reach, have also contributed to their appeal.

A typical video conferencing system digitally compresses and transfers audio-video content in real time over a network. Figure 6.2 illustrates the components of a standard video conferencing system.

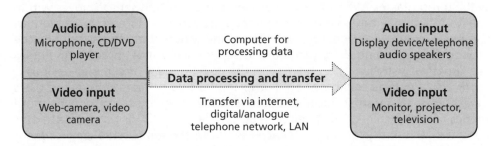

Figure 6.2 Components of a standard video conferencing system

A wide range of video conferencing tools are currently available in the market. Table 6.3 lists some of the tools available on the market and gives a description of each.

Impact on management practices

It is evident that the absorption of technologies offers competitive advantages to organisations by enhancing their processes and offering new alternatives (video conferencing) to traditional processes (face-to-face meetings). Environmental factors including the ever increasing globalisation of the industry and the fierce economic climate, combined with technological factors such as the progression of technology infrastructures and applications over the past decade, as well as the growing cultural dependence on technology in everyday activities, have called for a need now, more so than ever, to be proactive in the seeking of new initiatives in order to remain competitive (Henderson and Ruikar, 2010).

Management practices have in the past blamed the technology when desired results (e.g. return on investment) did not materialise. This, combined with the difficulty of quantifying the impact of technology within the construction industry due to the degree of fragmentation, has contributed to the slower uptake of new technologies in construction compared to other industrial sectors. There is, however, evidence that the future of construction is one that embraces technological advances and therefore the time has come where a strategic adoption of technology is required in order to extract its full potential.

Table 6.3 List of video conferencing tools

Video conferencing tools	Description
WebEx	WebEx is a web-based conferencing tool that offers multi-party collaboration solutions for online meetings, remote support, webinars and other online events. WebEx combines desktop sharing through a web browser with phone conferencing and video, so everyone sees the shared document while they talk.
Elluminate Live	Elluminate Live is a web-based conferencing tool that enables virtual meetings to take place in virtual rooms. The Elluminate Live software includes visual aids such as whiteboard, enables application sharing and file transfer, and allows the meeting 'moderator' to record meetings. The system is designed for use on all computers that have Java installed.
Skype	Skype's group video calling facility enables group video calls between three or more people (up to a maximum of 10). At least one person on the call needs a group video calling subscription. Call conference participants require a webcam, a high-speed broadband Internet connection and the latest version of Skype for Windows installed on their computers.
Vtok	The Vtok app enables free video and voice calls to Google contacts over a Wi-Fi or 3G network. It is supported on iPhone, iPad, iPod touch and some Android devices.
Adobe Connect Now	Adobe Connect Now is a video conferencing feature that is part of the free Acrobat suite that allows users to share and store files, provides mobile access, online PDF creation functionality and online office application add-ons. Adobe Connect Now is compatible with both Mac and PC requiring no downloads.

In order to improve technology implementation strategies, improvements are needed in addressing behavioural and emotional concerns. To have the greatest positive impact on overcoming these issues, a minimisation of the initial uncertainty surrounding change needs to be made. This is due to its evident link with increasing the levels of emotional distress of change participants (Henderson and Ruikar, 2010). To accomplish this, two-way communication, education and training, and understanding of the change rationale, should all be improved. These aspects can all be enhanced through focusing on increasing the levels of involvement of all participants (internal and external) at every stage of the process. Thus, the degree to which successful technology implementation is achieved ultimately depends on the extent to which changes are planned, managed and evaluated effectively. It is therefore not so much a technological issue as it is a human behavioural one.

Greater adaptability and acceptance of change can be created through a multi-dimensional approach requiring the effective leadership of; the process, technology, people and culture. This is due to the reduction of emotional distress throughout the change process. The extent to which this has been achieved should be evident in the degree to which the transition of employees from being 'change intolerant' (the late majority and laggards) towards becoming 'change embracing' (innovators and early majority) is observed. If this can become normality at all levels of an organisation, it is expected that improved levels of adoption can be experienced, thus leading to a reduction in the extent to which construction is perceived to be 'lagging behind' that of other industrial sectors.

An effective collaboration technology strategy is one that is 'change embracing', dynamic (responsive) and closely supporting its owner's strategic business goals and also those of its external collaborators. This is especially important in technology supported collaborative environments (i.e. construction projects) that force organisations to work beyond their traditional 'internal' boundaries. A key consideration for any strategy is therefore to consider not simply its internal environment, but also the external environment.

Conclusion

All of the tools reviewed in this chapter can help to facilitate collaboration and integrated working. They can also make a significant impact on the ability to work in geographically dispersed teams and geographically remote locations. From the design manager's perspective these collaborative technologies may be used alongside more 'traditional' forms of interaction, such as face to face meetings and workshops held in one geographical place. How these tools are taken up by design managers is addressed in Chapter 7.

7

VISUALISATION AND BIM

The ability of a wide range of collaborators to visualise the design, or rather the development of the design, is crucial to effective and efficient working. It is also crucial to the generation and delivery of value for the client. In this chapter, emphasis is on visualisation and BIM for design managers, with two case studies helping to identify some of the challenges and opportunities associated with visualisation and BIM.

Opportunities and challenges

Technological developments are continually evolving and creating new challenges and opportunities for all those involved in construction. These developments map very closely with advances in processing power of computers, increases in storage capacity, the graphics and image processing capability, and among other things, improvements in network connectivity. With these advances, it is possible to develop (2D and 3D CAD drawings), visualise (solid modelling), experience (via immersive technologies) and manage (via projects extranets and/or BIM) complex designs. Figure 7.1 illustrates how the evolution of the technology, impacts on computer outputs and user experience/interaction. For instance, advances in display systems means that users can now have a totally immersive experience in high resolution and full colour.

Even though technological developments create new opportunities, they generate new challenges. For example, now the move is towards a 'single' shared environment, where organisations can work collaboratively, through lifecycle, to realise the project. This is a shift from the traditional practices of working within professional and organisational boundaries (in silos), to an integrated approach where disparate systems are linked so that data, information and knowledge are shared without the need for re-input. To realise the total integration dream of single models, in the first instance, problems associated with discrete systems need to be fully understood and addressed. Often these problems occur because of inadequate capture and processing of project needs: 'silo' mentality of project team members; fragmentation of stages in the project's lifecycle; poor coordination of project information; lack of lifecycle analysis of projects; development of sub-optimal design solutions; inefficiencies (rework, delays, cost); and inadequacies in the communication of design rationale/intent. In addition, it is equally crucial for organisations to realise that total

CAD
From drafting ➔ Design ➔ Modeling and simulation

Computer graphics and image processing
From 2D ➔ 3D ➔ 4D ➔ .... nd (BIM)
From single sources ➔ Multiple sources of spatial data (GIS)
From non-real time ➔ Animation ➔ Real time
From wireframe objects ➔ Realistic renderings ➔ Virtual environment

Display systems
From cathode ray tubes ➔ Limited field of view projection systems ➔
    Head/eye tracked ➔ Helmet displays
From low resolution ➔ High resolution
From monochrome ➔ Full colour
From observer ➔ Participant

System archiectures and network
From mainframes ➔ Mini computers
From PCs and distributed processing ➔ Local area networks ➔ Wide area networks ➔ Internet

Figure 7.1 Evolution of technology and its impact on computer outputs and user experiences

integration requires a vision that extends beyond the traditional organisational boundaries to that which is shared by the collaborators within the design team.

A vision that values 'collective' knowledge, embraces technological advancements and aligns business goals, processes, technologies and people skills; at intra- and inter-organisational levels. These emerging requirements require innovative approaches to resolve deep rooted challenges. For example, companies now use visualisation capability of specialised tools to push 'creative' boundaries and generate 'virtually' real facilities that detect imminent management challenges (e.g. due to design clashes) that cause delays and disputes. One such example of how Arup used visualisation tools innovatively in the design process to develop and manage a unique, iconic, structure is described in Case study 7A.

Case study 7A

Visualisation tools in the design process

Chiara Tuffanelli, Arup, London, UK.

Choice of visualisation tools

Nowadays we are overwhelmed by a multitude of digital tools relevant to modern architectural and engineering practices. This considerable number, together with the increasing power of technology, can easily create an illusionary confidence that everything can be done easily, quickly and with just a 'click of the mouse'. That is not always the case; therefore, when a new project begins it is essential to make a careful and considered choice of tools to be deployed within the design team. The selection of these digital devices (e.g. visualisation tools) remains crucial throughout the entire design process, from the initial concept stage through to detail designs, tender, construction and occupation, right up to demolition or end-of-life stages. The choice of the best digital tools for production deliveries, for visualising and communicating ideas and for exchanging data results between the different parties involved in a specific design, will depend on a multitude of factors, a few of which are summarised in this case study.

Project topology and scale

Both the project typology and the scale of the design can be the main drivers for selecting proper design and visualisation tools for the design team. A landscape or urban planning design will probably have different production and visualisation requirements from a façade connection detail design, for example (Figure 7.2).

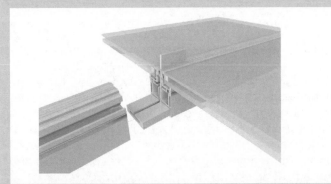

Left: A roof light purlin/gasket connection used during the design development phase in order to share geometrical concerns with the client.
Right: A high-rise building solar exposure analysis expressed in a 3D diagram.

Figure 7.2 Visualising different data at different scales
© Arup

A standard high rise building may be characterised by repetition of elements that might speed up the delivery production process, while a complex geometrical building or an elaborated artwork might be described by a necessarily high number of diverse geometrical elements, and hence may require specific investigative digital (visualisation) tools.

In general, the more complex the design is, the more the 3D model assumes a fundamental role for communicating ideas, concerns and data results between the designers, clients and construction partners. As a matter of fact, the final issue of the design is increasingly delivered as a 3D model package together with relatively few 2D drawings. Ideally the 3D model should be used in all design stages, the earlier the better, and typically it should mature together with the progression of the design. It can incorporate information from early stage hand sketches, diagrams and spreadsheets; and continuously be affected by the design choices made by different parties involved. At the same time, the 3D model will constitute a fundamental means for visualising and exchanging information and consequently influencing the evolution and progression of the design.

Design team legacy systems and expertise

Another key factor for choosing the correct digital tools for production, delivery and communication of the design is the analysis of the design team members' legacy systems (i.e. hardware and software components) and expertise. Not all of the design team members are necessarily a part of the same company; therefore they may not have access to compatible and/or same software packages. Also, there is the possibility that they may have compatible software, but varying degrees of competence in operating the software. It is therefore, essential to deliberate whether training is necessary and decide whether it should be part of the long-term investment plan.

Tools availability

Once the design team composition is clear, the focal point shifts to the company policies for the use of digital tools. Are the chosen software packages easily accessible? Could there be a real necessity for investing and purchasing a different software package? Is it worth investing in newer technologies? In cases where the use of diverse software packages is inevitable, the essential question to ask is; what type of common file format will enable a seamless exchange of data between all parties involved? Before arriving at a conclusion, many factors such as accessibility, time and costs need to be considered for achieving the 'ideal' solution.

Previous experience

The design team members' previous experience with a similar type of project is another crucial aspect to be taken into account. What software has been used in the past? Have there been any issues in communication, in the workflow, or in exchanging data between people involved? And if so, are there any particular tools that could avoid certain mistakes reoccuring?

Interaction and integration

Another significant aspect is the feasibility of collaborative work. At this point, the questions to ask are as follows. What level of collaboration is required to realise the design? Is everyone involved based in the same city/country/continent? Is video conferencing necessary? What communication tools are available to disparate team members?

Once the basic choices have been made and a good management protocol is followed, digital tools can help a great deal by accelerating the process of communication between different people and different technologies involved. The parametric tools largely in use for complex geometrical designs can be an example. A parametric model can adapt in real time to the continuously changing requirements and therefore it can be modified directly during the meetings themselves. Figure 7.3 is an example of a parametric model used for generating the geometry of the ArcelorMittal Orbit sculpture for the London 2012 Olympic Games.

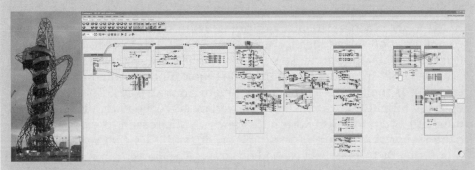

Figure 7.3 A parametric model of the ArcelorMittal Orbit project
© Arup

Representation style

Once the executive and technical aspects have been contemplated, the team can concentrate on the method and style of communication itself. It is quite common to use hand sketches together with diagrammatic 3D model renderings at the initial stage of the design. Rough 3D objects can be quickly modelled in a variety of 3D software and are used for generating concept images and data. Once the design progresses, a further level of detail can be then achieved. Photorealistic renderings can be produced in order to express thorough product characteristics such as, for example, materials. Often though, the renderings are deliberately kept at a diagrammatic level until the final stages of the design in order to avoid contradictions with technical specifications.

Along with 2D images and 3D models, many other means of expression such as videos, multimedia reports, websites and interactive reports are increasingly used. For example, Figure 7.4 is an extract of a video of a diagrammatic simulation of the build up of a Kalzip roof system.

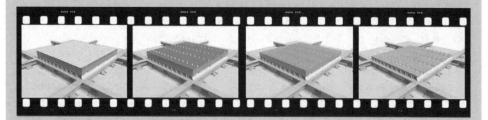

Figure 7.4 Video screenshots of a Kalzip roof system build up
© Arup

At present, a great deal of effort is being made to achieve the best communication techniques for simulating the 3D world onto a 2D surface (such as paper, computer screens or mobile devices). One of the best achievements in recent years for communicating design has been the creation of the Adobe Portable Document Format (PDF). This is a widely used file format that can be easily viewed by everyone without the need to purchase additional software packages. For this reason, besides standard paper reports, Arup are increasingly producing iPDFs that are interactive reports in PDF format, best viewed on screen. The viewer can easily navigate through the report in a similar way to a website, with button fields that guide users through the different pages of the report. Arup's are now broadening the application of 3D PDFs in building design.

'The tall tree and the eye'

The example discussed here is that of 'The tall tree and the eye' sculpture by Anish Kapoor (Figure 7.6), where Arup's engineering tradition has been combined with innovative outlook on structure, form and aesthetics. This sculpture is characterised by a very intricate geometrical design and therefore has required appropriate 3D visualisation tools throughout the project history. The design complexity together with a very tight programme schedule entailed a choice of proper visualisation tools that could, at the same time, be flexible, precise and accelerate the process.

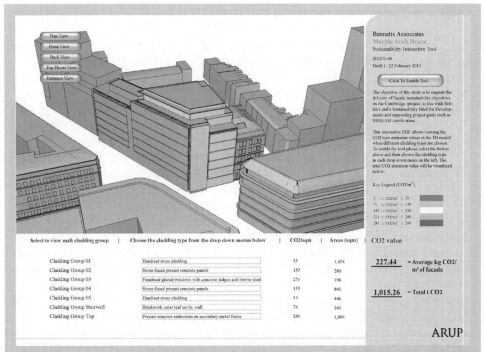

The PDF format can now incorporate a 3D model that enables the user to navigate inside a 3D space without the use of complicated and expensive 3D modelling packages. The image at the side represents an example of a 3D PDF where CO_2 datasets, input from a spreadsheet, can be visualised real time onto the 3D model itself.

Figure 7.5 PDF documents with embedded 3D models
© Arup

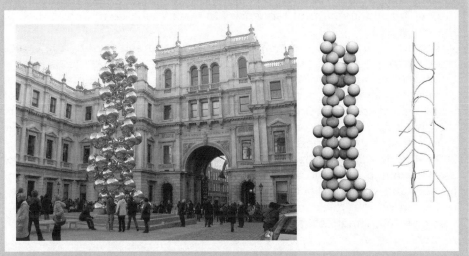

Figure 7.6 Anish Kapoor's 'The tall tree and the eye' sculpture in the Royal Academy of Arts courtyard, London, 2009
© Arup

Description

73 mirror polished stainless steel spheres are stacked to a height of approximately 14m and a width of 5m. Each sphere has approximately 1m diameter and a maximum wall thickness of 2mm. The spheres are positioned in such a way that they completely hide the structure contained within. This internal structure consists of three carbonated steel masts linked together by curved bracing elements and is connected to a steel base frame which rests at ground. Due to its particular geometry, the main challenge in designing this sculpture has been to achieve the invisibility of all structural elements together with the least possible amount of structure.

The 3D model

As discussed above, this elaborated sculpture required the use of a 3D computer model since the early stage of the design due to its geometrical complexity. 2D drawings would not be exhaustive and would not easily express all of the design's sophistication. Furthermore, the small team of designers and engineers involved needed a digital tool that could easily and quickly exchange data within different software packages. The choice was narrowed to the Rhino software (http://www.rhino3d.com/) as it can offer flexibility with 3D complex geometrical modelling. At the same time, it can provide good standards for renderings production and parametric properties.

The model has developed throughout all design phases until the completion stage. The initial 3D model was based on the analysis of regular and irregular sphere packing systems in search of a possible arrangement of spheres that would hide internal columns and would respect the necessity of keeping all the points of tangencies to a minimal area throughout all of the sculpture.

The artistic intent of the sculpture is deeply characterised by the choice of a mirror effect so to create the appearance of weightless floating bubbles rising into the sky. For this reason studies on reflection properties of mirror spheres have been carried out in order to enhance the control of visual impact that the sculpture would have on future observers. Through the analysis of basic convex mirror properties, the study of the hyperbolic space (Arnold and Rogness, 2008) and principles of inversion (Kalajdzievski, 2008) the design team demonstrated the reliability of the rendering tool used (Flamingo, a Plug-in for Rhino). As a result, the numerous 3D computer models and computer-generated images aided visualising the effect of reflection on multiple tangent spheres. Interactive photo-realistic renderings generated images that have been then verified on site as very close to reality (Figure 7.7).

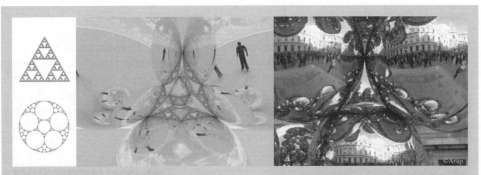

Left: The Sierpinski and Apollonian gasket on the left, a tetrahedron of rendered spheres and a photograph of the sculpture express the reliability of the rendering tool used.
Right: The reflection of perfectly reflective tangent spheres into each other generates an infinite fractal pattern.

Figure 7.7 Interactive photo-realistic renderings
        © Arup

The parametric model

The form finding process of the conceptual design had to integrate aesthetic requirements, but most of all geometric and structural ones. Once a 3D layout succeeded aesthetically, both from a reflection and a volumetric point of view, it would be exported into a structural analysis software Oasys GSA (a structural analysis software developed by Arup) to verify local and global stability of the sculpture. Finite Element Analysis models would be elaborated based on various mesh sizes and element types. Investigations of visual impact due to sphere deflections would also be analysed in order to control any visual change in the reflections. All structural analysis results would then be used as feedback for further changes to the 3D model that, in a back and forth process, would generate further considerations resulting in a new model to be once more analysed structurally.

As a result, the design of this sculpture required a 3D model that could adapt and change quickly, accordingly to the design and structural progress. The need for a parametric model could be achieved either through implicit or explicit history tools tightly integrated with Rhino's 3D modelling tool.

The best solution for this project has been provided by the explicit history tool named Grasshopper,[1] a graphical algorithm editor where parameters are assigned and linked through several components that control the 3D model. The best characteristic of this tool has been the fact that, differently from the implicit history tools, it could provide an immediate visual feedback and full control on each single component and stage of the process created by the user. Whenever one of the parameters would change, the entire model would adjust consequently to suit the initial requirements. In this way, any geometrical variation required from aesthetic or structural reasons could be rapidly exported into analysis models for structural tests.

Conclusions

The example discussed here illustrates how the use of new computational 'visualisation' technologies can help the coordination of multiple design aspects, from the most theoretical ones, such as fulfilling the artistic concept, to the most practical ones, such as complying with structural performance and manufacturing requirements. These tools can extensively facilitate as well the communication between the client, the architects, the engineers and the manufacturers throughout the entire design process.

Figure 7.8 The 'Tall tree and the eye' covered by a protective film that was removed on site

For the 'Tall tree and the eye' sculpture (Figure 7.8) the 3D parametric tool has been very useful for communication through renderings and 3D prints. But, once the geometrical model was built and finalised, it has also been valuable for extracting polar coordinates of tangent points between each single sphere and exporting them into a spreadsheet. This would allow an easy exchange of information on this stable and yet light structure between Arup, the fabricator and the contractor. Through the information of the spreadsheet, it was possible to control and rebuild easily the 3D model from scratch, with any software accessible to the manufacturers. For this reason, it was, of course, beneficial also for cross checking that any information issued was coherent with the design. Therefore it is crucial for the design team to examine and deliberate as soon as possible on the type of digital tools employed in the project as they might help or complicate, if misused, the design process in all of its phases.

In conclusion, it is further advised that the reliability of tools to be used is verified at the onset. The result of misuse or non-reliable tools can highly affect the design and as a consequence, the total time and costs of the project.

BIM for design managers

As already discussed in Chapter 6, BIM is a digital representation of the physical and functional characteristics of a facility (BuildingSMART, 2012). It enables complex building and project information to be modelled into a single, virtual model that represents all elements of the building or project. However, BIM is not just a single tool. According to Eastman *et al.* (2011), BIM is a modelling technology and associated set of processes to produce, communicate, and analyse building models. The model comprises of a collection of intelligent components that know what they are and can be associated with computable graphic and data attributes, and parametric rules. This can be better understood with the example of a simple building component such as a window. During a design or a construction process, the information of a window is stored in several different documents. Examples include:

1. architectural design drawings (conceptual designs, scheme designs, detail design, etc.);
2. references to the window in a number of other design drawings (e.g. structural, services, etc.);
3. window schedule;
4. cost schedule;
5. project plan that sequences the installation of the window;
6. the 'actual' product listed in the manufacturer's product list;
7. the supplier who delivers it as per the project plan, etc.

All the above information is interrelated and even if one attribute changes (e.g. an increase in opening size requiring larger windows) it affects the other related pieces of information. Without intelligent components, if any information changes do occur, the relevant personnel have to make changes to each and every individual instance of that component. This practice is susceptible to human error. If, however, intelligence is embedded into the component(s) (using the capability that BIM tools offer) it would result in automatic updating of all the related information. Thus, problems associated with working off the wrong or incorrect information and the resultant impact on project costs are greatly reduced.

The coordination and computational properties of the BIM data is what separates it from the traditional 2D or 3D models. A traditional 2D drawing is simply a geometric representation of the facility, whereas in BIM, objects are parametric. In a parametric design, instead of designing a separate instance of the building component (say wall), the designers define classes (or a family) and a set of relations and rules to control the parameters by which the component instances can be generated (Eastman *et al.*, 2011). These parameters assigned to the building components allow the objects to interact with each other and as a result affect the geometry of the related objects. For instance, if the single door is replaced by a double door, the 'opening' in the wall will be automatically increased (in every instance) to accommodate this change. Thus, parametric building information models are characterised by:

- Components that include data that describe how they behave, as needed for analyses and work processes, e.g. takeoff, specification, and energy analysis.
- Consistent and non-redundant data such that changes to component data are represented in all views of the component.
- Coordinated data such that all views of a model are represented in a coordinated way.

(Eastman *et al.*, 2011)

Impact of BIM on design-construct activities

Given its through lifecycle capability, there is no doubt that BIM impacts on upstream and downstream design activities. Any changes to the activities would pose challenges to the design teams including the design manager. For example, from a design perspective, BIM provides simulation and analysis opportunities to the design teams. Design teams can perform BIM simulations such as the 'live' monitoring of buildings for energy and water use to ensure that the design meets sustainable building targets (e.g. BREEAM – BRE Environmental Assessment Method). Furthermore, the design teams will be able to share this information with other team members, so that their processes are faster, transparent and efficient.

Structural engineering teams can utilise BIM data for costing, financial planning and cash flow generation. Any changes to the original design can automatically be reflected in the cost estimates, helping design teams to take informed decisions, and perform 'what if' analysis on the BIM models to assess what impact any changes to the structure of the building have on the material costs. The design drawings production teams would then be able to deliver fully coordinated drawings indicating design changes to the relevant project teams saving time and cost of production. This multi-dimensional, collaborative capability breaks traditional practices of working in silos by enabling a shared and cooperative approach. Thus, with the help of BIM tools, relevant design team members can collectively coordinate, control and monitor information such as, specifications, plans for demolition, safety reviews and coordination details, allowing the contractors to foresee unexpected scenarios during construction (Eastman *et al.*, 2011). From a client's perspective, BIM provides clients with the opportunity to witness multi-dimensional models during early stages of design giving them the opportunity to comment on and/or make changes during early stages of design. These changes would automatically 'show' what other aspects of the building design are affected as a result of the change. For example, if the client team decides to change the position of a staircase (which may ordinarily seem as a minor change to a client who is not particularly experienced in construction projects) the decision would have several implications that are detrimental to the project. The change would lead to a 'ripple' of changes. For instance, the foundation pits would need to be re-dug in the changed location, the structural designers would need to verify whether the existing beams and columns can cope with the changes to the loads (due to unexpected human traffic), the respective engineers would need to check for clashes with the HVAC systems (heating, ventilation and air conditioning), electrical cabling, and so on and so forth. This 'change ripple' poses coordination challenges (re-sequencing and re-scheduling of activities impacted by the change), which inevitably lead to delays and disputes. A BIM environment can visually demonstrate the scale of impact of the proposed changes. This can reduce the risk of making costly mistakes or unacceptable scenarios during design and construction phases preventing serious project delays (Eastman *et al.*, 2011).

Post-completion of a project the BIM model also creates opportunities for the FM (Facilities Management) teams, by making it easier for them to access the as-built documents which improves their understanding of the building operations. The maintenance team can use BIM to budget for maintenance and schedule/sequence maintenance activities.

The as-built archive would also be useful to owners and occupiers of the facility. It not only improves their understanding of the building itself, but also provides them with vital information contained within (and about) the 'skeleton' and 'genetic make-up' of the building they occupy. This would be particularly useful in renovations, repairs, extensions and/or alterations that are planned in the future. Currently, occupants have little or no access to the original plans and drawings of an old building, unless alterations were made to the original building, in which case the documents may be acquired from the local council offices. This, however, is a laborious and time consuming process; the outcome of which is not always as desired.

It is quite clear that the benefits of BIM extend beyond the realms of traditional design-construction functions to include through lifecycle functions. In a BIM enabled environment, each project team's function generates inter-related knowledge sets, all of which originate from the 'single' model. Thus, the resulting BIM enabled workflows challenge traditional 'over the wall' operational workflows in which collaboration and communication mostly took place at the information handoff stages. These workflows, present new implementation and operational challenges to the project teams, requiring development of strategies that accommodate the emerging needs. The following case study presents an industrial perspective of implementing BIM in practice.

Case study 7B

BIM practice: an Arup perspective

Dan Clipsom, Structural Engineer and BIM Coordinator, Arup, Edinburgh, UK

The implementation of BIM across the UK construction industry presents both great opportunities and significant challenges. The basic philosophy of BIM is the application of cutting edge technology and new working practices to drive a more integrated, collaborative and information-centric approach to construction projects.

As of 2012, the construction industry in the UK is grappling with the widespread adoption of BIM in the face of rapidly increasing client demand. A whole-scale rethinking of the way construction projects are executed is underway, but the legacy of traditional contractual relationships and attitudes, along with tough economic times mean the challenges of implementation are many.

Challenges to BIM implementation

Recent experience of spearheading BIM on projects within Arup has shown that these challenges can be overcome by tackling them in manageable pieces and ensuring that equal focus is given to both process and tools, ensuring that BIM implementation becomes as much about people as it does about technology. The specifics of this can be broken down into three stages of a project: (1) planning; (2) executing; and (3) delivery. Each new stage represents a change in priority from the last, but builds on the momentum created. The following cases are cited from the perspective of design management in a multi-disciplinary project team.

Stage 1 – planning

The first stage is the most crucial for BIM, as the changes it requires from traditional working practice need a willing consensus from the team built around a clear and common understanding of what needs to be achieved.

A PFI (Private Finance Initiative) education project in Glasgow is a good example of how this can be achieved. In this case the project team consisted of designers, builders and operators integrated together from the beginning of the project bid. At a very early stage, a project BIM workshop was held with all parties present and an agenda biased towards the production of a 'BIM Execution Plan' for the project.

The workshop had three objectives: to educate the project team on the principles of BIM, to map out a collaborative workflow for the lifecycle of the project, to get agreement on a level of integration within the project team that goes beyond the minimum required by the contract.

This kind of planning process is driving new kinds of conversations within the team and generating a more collaborative attitude than has previously been the case. While the creation of an agreed BIM Execution Plan and its specific content is increasingly prescribed by clients and present in published guidance, it is the actual process of bringing the team together to produce it that is bringing about positive change.

Stage 2 – executing

A recent cultural development in Dundee serves as prime example of how BIM technology and process can come together to improve collaboration within a distributed design team during the execution of a project. The building in question is of a complex geometrical form with tightly defined functional and aesthetic requirements and a combination of onerous site constraints. The decision was made to coordinate the building through a central shared 3D geometry model that contained all the key aspects of the building and would serve as the basis for the decision-making process on key aspects of the engineering and architecture. This shared model far exceeds the basic requirement to exchange 2D drawings of the building, but there was an appreciation of the need for something better than a traditional approach to ensure the building was a success.

The 3D geometry became the language of collaboration on the project and in doing so drove efficiencies throughout the disciplines involved. The structural and building services disciplines within Arup started working in real time with each other's model data so coordination is now an on-going daily dialogue instead of a series of discrete interventions. Engineers and architects and specialist consultants are testing, simulating and analysing the form and function of the building based on shared information which is then in turn being used to asses impacts and manage the changes that are implemented.

The technology deployed in this case has driven a lot innovation but more crucially it has created a more open and honest atmosphere within the team and a shared sense of responsibility for getting things right.

Stage 3 – delivering

Stage 3 concerns the delivery of design information down the construction supply chain. This is perhaps the least developed of the three stages and one which is fraught with on-going technical challenges and legal issues still to be overcome.

Often the contractor and construction supply chain is not part of the BIM planning phase and so assumptions have to be made about which potential gains could be realised. The initial steps are small but momentum is building. Taking the structural engineering discipline in particular, several current projects have taken a step towards capitalising on Stages 1 and 2 above to deliver enhanced design information to a contractor in order to improve accuracy and efficiency within the construction process itself. A recent cultural project in Glasgow experimented with the delivery of 3D digital, setting out information for a complex curved roof structure directly to the steelwork fabricator. In return the fabricator issued a fully populated 3D steelwork model back to the design team which contained all the fine detail required for checking of the design.

Again, through the agreed exchange of information surpassing the minimum requirements of the contract, a more collaborative environment was created in which problems were identified and resolved earlier and more effectively than has previously been possible.

Lessons learned

To sum up the above cases, experience has shown that generating real and measureable benefits from BIM requires a structured approach to planning and managing a project team in order to create a more collaborative set of practices and attitudes. The level of challenge remaining will require a focused approach to individual projects, but each success builds towards the overall goal of bringing about a cultural change towards increased collaboration.

Conclusion: impact on design management practices

It is evident that if designs are to be developed and managed using BIM, it is not simply the design processes that will be impacted, but also the roles of the design managers that would be affected. These changes would require the design manager to:

- Have an 'understanding' of influencing and effecting the organisational changes brought about by BIM;
- Possess empathy and respect for *all* (existing and emerging) roles and disciplines within the multi-disciplinary design teams;
- Be competent in setting up and maintaining, BIM enabled design management systems, including their integration with cross-disciplines;
- As a minimum, possess a working knowledge of BIM while demonstrating a deep, strategic understanding of design management practices so that design value is realised and quality standards maintained;
- Possess skills to recognise 'emergent' trends and map these against existing design management processes to identify potential bottlenecks and any future opportunities; and
- (Above all) have a natural enthusiasm for learning and sharing knowledge to realise new opportunities.

As the use of BIM within the industry matures, simply possessing a working knowledge of BIM will not be sufficient. The design managers would not only have to: (a) continue to perform the traditional role of design coordination of multi-disciplinary design teams from inception to handover; but also (b) develop new skills needed to fulfil BIM enabled coordination. This would include new responsibilities that require the design manager to incorporate constructability and construction resources and methods into BIM, 'read' and utilise the model data to coordinate multiple project disciplines, monitor, record and report project performance, generate 'lessons learned' case study archives and keep abreast of BIM led advances.

Further reading

Comprehensive information on BIM can be found in the *BIM Handbook: a guide to building information modelling for owners, managers, designers, engineers, and contractors*, 2nd edn. (Eastman, C.M., Teicholz, P., Sachs, R. and Liston, K., 2011, Hoboken, NJ: John Wiley & Sons).

Note

1   Grasshopper is a Plug-in for Rhino Software, developed by David Rutten. Further information available from: http://www.grasshopper3d.com/.

8

COLLABORATIVE WORKING STRATEGIES

In Chapter 1 we outlined some of the fundamental issues associated with collaborative working. Now, as we approach the conclusion of the book, we return to this central theme and look more specifically at the strategies for collaborative working. As inferred throughout this book the role of the design manager, and to a certain extent the design integrator, may be influential in the adoption and effective use of new technologies to facilitate collaborative design management. These strategies need to be aligned to the business and the project portfolio, a central tenet of effective design management thinking.

Key components for collaborative working

Information and communication technologies (ICTs) are at the heart of communication and collaboration in construction projects, therefore any strategy for collaborative working should place technology at the heart of it. And at the heart of any technology are information and knowledge. Many organisations fail to realise that installing systems without due consideration to the strategic implications, would invariably result in 'knowledge leaks' and hence a loss of business value. To avoid ending up with software installations instead of comprehensive solutions to real business problems, organisations would need to adopt a strategic approach. A successful cross-organisational rollout is therefore more than simply buying and installing technology applications. The most effective strategy is the one that closely supports its owner's strategic business goals. The result is the ability to capitalise on the full potential of the technology investment, both in the short and long term.

Considering this, it is essential to develop short-term and long-term strategies for the effective deployment of technologies. The approach should be a holistic one that addresses every facet of an organisation and produces substantial changes not just from a *technology* perspective, but also from the internal organisation's perspective of its *management* practices, *people* (culture, attitudes, behaviour), and *processes*. Several research publications (e.g. Basu and Deshpande, 2004; Goolsby, 2001; IBM, 1999; and Kern *et al.*, 1998) and articles (e.g. Fuji Xerox, 2003; Larkin, 2003; and Emmett, 2002) indicate that people, processes and technology are the three key aspects that need to be considered for the successful implementation of technologies. Emmett states that together these three elements generate business value. However, he further states that 'the people, processes, and technology need a leader', just as 'an orchestra needs a conductor'. Emmett (2002) draws a parallel with the performance of an orchestra and states,

> 'in an orchestra...You've got musicians (people), musical scores (process), and musical instruments (technology). But without a conductor, they're more likely to produce noise than music. Even if everyone in the string section plays the right notes at a relatively similar tempo, creating a symphony requires more than following the sheet music.' Therefore '...An orchestra needs a conductor.'

The same analogy can be applied to the adoption and implementation of new and innovative technologies within construction companies. The 'conductor' in this case is the management. To successfully implement and use any new technology it requires management buy-in and belief in order to plan and drive policies and strategies. The adoption of any new and innovative technology within an organisation, department, or work group requires total commitment from the management (or group leader). It is important for the management to buy into the technology so that they can lead the business into successfully implementing and adopting the technology, i.e. the management needs to be e-ready. Thus a fourth category, 'management' is necessary. These four components are weighted equally and for the basis of an e-readiness model, VERDICT (an acronym for Verify End-user e-Readiness using a Diagnostic Tool) that was developed by Ruikar (2004), to spring board organisations into achieving readiness. A company cannot be e-ready if it satisfies the requirements of just one category and not the others. For example, even if management, processes and people are e-ready, the fact that the technology infrastructure is inadequate will affect the overall e-readiness of the organisation. This example indicates that an organisation would need to address its technology issues in order to be e-ready. Drawing from the orchestra analogy, 'a memorable symphony performance doesn't happen when the players just assemble with their instruments and scores.' and, 'the orchestra with the most violinists isn't necessarily going to sound the best.' Similarly, all four categories – management, processes, people and technology – need to work hand-in-hand and symbiotically. It is also important to recognise that today's technology tools drive (and in some cases 'force') collaborative working, which extends beyond organisational boundaries to those of other supply chain collaborators. Thus, any strategy that is developed needs to take into account both, the internal and external environments (Figure 8.1).

Central to a profitable project portfolio is effective collaboration. One main challenge of collaborating with external parties is due the differences in working practices. For example, SMEs (Small and Medium-sized Enterprises) do not necessarily have access to the same level of resources as larger organisations. In SMEs, staff are often overloaded with multiple tasks, which impacts on the organisational performance and efficiency (Major and Cordey-Hayes, 2000). This consequently impacts on the collaborative productivity. Thus, recognising the differences in working practices among collaborating organisations is the

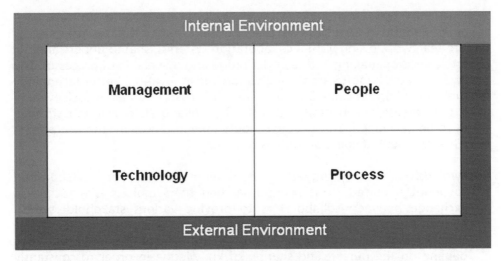

Figure 8.1 Aspects to consider when defining strategies for collaborative working

first step towards attaining value from it. Profitable organisations derive value from the collaboration, by devising strategic collaborative agreements that:

• recognise the various characteristics of collaboration (Grieves, 2000);
• encourage collaborative decision making (Mudambi and Helper, 1998; Cardell, 2002);
• recognise the need to actively manage multi-stakeholder interests (Boddy *et al.*, 1998); and
• share benefits equitably (Cox, 2001).

Prime contracting is an example of a strategic collaborative agreement. Prime contracting is used to ensure pain-gain-sharing mechanisms (OGC, 2003). The Andover North redevelopment project was one of the earliest prime contracting projects that included a partnering charter, joint project bank accounts, and collaborative IT, among other things (Gohil, 2010). It used collaborative charters to ensure that the project promotes aspects of collaborative working that include:

• focusing the whole team on delivery;
• sharing risk equally;
• managing risk rather than transferring it; and
• continually assessing cost, time and quality.

Formalisation of strategic collaboration clearly sets out expectations of what needs to be done to help establish a 'common' standard practice (Daugherty *et al.*, 2006). The benefit of such an arrangement is the elimination of ambiguity because of clearer priorities. Lu *et al.* (2007) divide the benefits of collaborative working into two categories; (1) teamwork paybacks; and (2) task-work paybacks. Benefits in teamwork are due to better communications, prospects of remote teamwork (using virtual collaboration systems), shared understanding, collective decision making and clear definition and ownership of processes. The task-work related benefits are through improved innovation, better technology integration, enhanced value and lower resource costs (i.e. shared costs).

To fully realise the benefits, an effective collaborative working strategy needs to be developed. This strategy should take into account the characteristics (Grieves, 2000) of collaboration as follows:

- Symbiosis, or synergy, suggesting the different organisations collaborating to adopt a united front for mutual benefit. Symbiosis is a reciprocal exchange between collaborators to provide various stakeholders with added value.
- Development of knowledge intensive firms through networks. They depend on the pursuit and sharing of knowledge in order to constantly innovate through multidisciplinary expertise and collaborative learning.
- The strategic management of a group of independent collaborators who gradually and collectively become interdependent, and
- An organisation culture that is shaped by knowledge sharing and trust between the various stakeholders.

Profitable organisations appreciate and value the link between effective management of their project portfolio (external environment) and effective management of the office environment (internal environment). One of the key facets of the literature relating to architectural management and design management is the recognition of the link between the successful management of design within the office and the successful management of design within each and every project. It is the synergy between the project portfolio and the office that influences the profitability of the business.

The core aspects of management, people, technology and process are discussed in more detail below.

Management

Management is an important factor that leads and governs the adoption, implementation and use of technology within organisations through the careful orchestration of business strategies in order to derive definite business benefits (Ruikar et al., 2006a). In the context of collaborative working, this can be achieved by: (1) defining specific business strategies for technology enabled collaborative working; and (2) ensuring that adequate resources (funds, time and man-power) are available. It is important that the management does not lose sight of its ultimate vision and aim of using the technology; i.e. to derive business value from the collaboration.

Management buy-in is an important aspect that can influence the successful implementation and adoption of collaborative technologies within a construction organisation (Ruikar et al., 2006a). Senior managers can authorise investigations into all aspects of current activities to identify areas where improvements can be achieved by changing to the new systems (e.g. BIM). However, management should endorse the technology only after investigating the technology for its overall capability and scope in addressing the 'collaborative' business objectives. For example, management should examine whether the technology has previously successfully nurtured collaboration in construction and/or other industries. It is equally to monitor its performance and capture lessons learned whether favourable, or otherwise. In the absence of lessons learned reports, it is worthwhile identifying and investigating the possible risks and taking adequate measures for minimising the risks and maximising the rewards. It is imperative that technology adoption will bring about changes (in existing communication channels and the outcomes of communication). Management needs to carefully consider different aspects of how these changes will be brought about and managed. Questions to consider would include. (Why) is the change necessary? Who will be impacted by it both, externally and internally? In what way? What can be done to prepare for it? How will it be implemented and monitored? And so on. These aspects are highlighted in Figure 8.2.

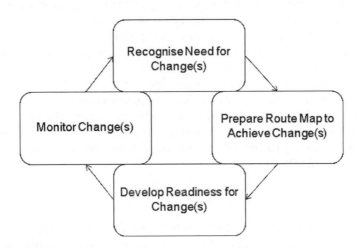

Figure 8.2 Aspects of change management

Careful consideration of these change aspects would help to ensure that value is derived.

People

The people factor accounts for the social and cultural aspects of an organisation. The people factor is an important factor that defines the success or failure of collaboration at all levels. This aspect takes into account the attitudes, outlooks, and feelings of staff within an organisation towards changes brought about by any new initiatives. People make organisations and are vital for its success. No matter how carefully the management has geared the business to nurture collaboration and derive value from it, it is less likely to fulfil its full potential if the people are not willing or cooperative. Thus, if organisations were to adopt new technologies (e.g. BIM tools) that are likely to impact on how people collaborate, the first step would be to understand the intricacies of the collaboration itself (why, when and how it takes place) and then devise measures (policies, processes) that ensure that it is fostered, essentially a 'bottom-up' approach. A way forward might be to ensure that the people have the appropriate skills and competencies, functional expertise, the right attitudes (towards cooperation and teamwork), a positive mindset, and the culture to adapt and adopt. This implies the need for managers to understand their co-workers and regularly review staff skills and competencies; a process that can be assisted by a carefully thought through continuing professional development (CPD) strategy for both the business and the individuals.

Collaborative working is a highly social phenomenon and stakeholder collaboration is often insufficiently informed by social science concepts (relating to motivation, team building, organisational culture, etc.) that are central to the understanding of collaboration between organisations as well as individuals (Shelbourn et al., 2006). Normally, the basis of collaboration is a partnership agreement prepared in view of protecting oneself from opportunism (Yli-Renko et al., 2001); although another more sustainable and positive use can be in creating and maintaining a long term relationship between the parties (Frankel et al., 1996). Given human bounded rationality, specifically in long-term agreements, the uncertainties arising over a longer period of time may lead to incomplete agreements (Hart and Moore, 1999; Maskin and Tirole, 1999). Hence, a partnership agreement should ideally cover soft aspects of management, referring to the development and management of relationship capital (Cullen et al., 2000).

Process

Process means a practice, a series of actions or workflows undertaken for a specific purpose i.e. fostering and deriving value from collaboration. It includes working rules, ethics, and procedures, within and between organisations. It is an important factor to consider, as any change directly affects intra and inter-organisations' processes. For example, when implementing a technology that impacts on collaboration processes (either incrementally or radically), it is imperative to ensure that the new technology either complements existing processes or that the existing processes are flexible enough to accommodate the technology. To enhance value from the collaboration (through increased transparency, accountability and cooperation; improved knowledge sharing, 'decision' audit trails, reduced response time; and improved integration of activities across the supply chain), organisations would need to examine and map their existing processes. This will help in identifying bottlenecks and devising measures to remove process inefficiencies.

The process related change that a technology can bring about is, according to (Laudon and Laudon, 2002) four-fold:

- Process automation, which is the most common form of change where organisations use computers to speed up the performance and efficiency of existing tasks and functions.
- Rationalisation of procedures, which involves the streamlining of standard operating procedures, eliminating obvious bottlenecks, so that automation makes operating procedures more efficient.
- Business process re-engineering, which is a powerful form of organisational change in which business processes are analysed, simplified and redesigned.
- Paradigm shift, which involves rethinking the nature of the business and the nature of the organisation itself.

Process change of any nature carries its own rewards and risks. Process automation and rationalisation of procedures are relatively slow moving and slow changing business strategies that present modest returns, involving lower risks. By comparison, the much faster and more comprehensive change brought about by process re-engineering and paradigm shifts potentially carries higher rewards and also higher risks.

Processes within an organisation can be divided into operational, organisational and strategic processes (Maleyeff, 2006; Huq, 2005; and Das and Teng, 2000). In project scenarios, due to the inherent complexity of the collaborating organisations, the organisational boundaries of the collaborating stakeholders are often fluid (Ritter and Gemeunden, 2003). The failure to formally describe collaborative processes can lead to ambiguity and inconsistency between the collaborators. Projects like Planning and Implementation of Effective Collaboration within Construction (PIECC) (Shelbourn et al., 2006) break down the ambiguities by developing associated processes for collaborating organisations.

Technology

The ubiquitous use of technology means that it is a crucial factor that needs attention. The technology factor covers all aspects of ICTs, which include both hardware and software; and their availability (within or between companies, departments and/or work units). Also important are the aspects related to the performance of the technology – thus, even if the technology infrastructure is adequate and available, it is of little or no consequence, if it cannot efficiently perform the strategic business functions. For example, an end user company may regularly upgrade their internal software (modelling software) without upgrading the network connectivity. In such an instance, the modelling technicians may be able to develop high end simulation models, but are not able to share the files because the system is not equipped to handle such a task. This affects the effectiveness of the collaboration, as the problem in this case is not simply confined to that individual company, but extends to its collaborators. Current collaboration technologies, such as extranets and BIM, allow project teams to communicate and exchange project knowledge in collaborative settings. Thus, if one company in the chain is ill equipped, it may have an adverse effect on the entire supply chain. This is an important issue that needs to be considered when developing a strategy for collaborative working in a sector like construction which is project centric.

Also important is the recognition that technology is capable of coordinating different activities within and across organisations, and also across industries (Laudon and Laudon, 2002). Processes can be made more efficient and streamlined by removing any obvious bottlenecks. This requires effort to observe the flow of work to identify where and when the bottlenecks occur and their impact on work flow. This is a task ideally suited to the design manager. However, these benefits cannot be realised if an adequate technology infrastructure is not in place. Organisations that aim to use collaborative technology systems (BIM) should in the very least have the basic infrastructure needed to operate the systems. Furthermore, it is important to establish if companies have had any previous experience of using collaborative technologies (e.g. project extranets), as this experience is invaluable when migrating to the next level (e.g. single 'project' model). This means that some attempt must be made to understand the potential project contributors before they are appointed, not afterwards (see Emmitt, 2010).

As discussed earlier in this chapter, the implementation of any new technology for achieving business targets necessitates changes to an organisation, its current practices, systems, human resources, processes and workflows. To enable this, the right strategies and implementation plans need to be developed, communicated, implemented and regularly monitored. This is not easy, so issues such as 'buy in', defining a strategy, selecting a system, developing a training programme, defining operating procedures, modifying organisational structures, and reviewing and extending uses need to be considered as part of the business strategy. While there have been noteworthy developments in addressing the 'readiness' needs of organisations to adopt emerging systems (Ruikar et al., 2006a) there is very little evidence to suggest that construction companies develop long term technology strategies to address their emerging business needs. Identification of the readiness needs is the first step towards ensuring that the organisation has the basic ingredients that are necessary for collaborating effectively. The VERDICT tool developed by Ruikar in 2004, helps organisations to successfully assess and achieve e-readiness.

Early approaches were often based on an overriding assumption that new technologies and process initiatives could be painted onto the fabric of the construction sector and implemented in the same way as in other industry sectors. These focused on 'new ways of working' and have tended to ignore the significant structural changes that have characterised organisations within the construction sector. Established contracting organisations have become progressively more removed from the physical work of construction, choosing to concentrate on management and coordination functions (ILO, 2001). This change towards being a service oriented industry has made it crucial that organisations develop long term strategies to remain competitive. Collaboration led technology adoption can no longer be viewed by organisations from an isolated technology centric stance. Instead, it must be integrated into the 'collaborative' business, from strategy, to implementation, and then on to measurement, analysis, reflection and feedback. Whatever the business imperative, if the change does not generate 'new' business value or fit into the overall business architecture as a key competence required by the enterprise, it is of little consequence. This ability is particularly critical with the speed and volume of emergent disruptive technologies. The most successful businesses will be adept at drawing technology landscapes that put emerging technologies into the context of their corporate strategies (collaborative working); indeed technology strategy should appropriately shape the corporate strategy. This will help identify both the potential benefits and any adoption or technology issues, such as complexities or lack of standards that might interfere with building business value (Ruikar *et al.*, 2006b) An effective strategy will be that which considers questions that fall into three interactive domains of 'Vision – Reality – Reflection'. Where:

- vision will focus on the future aspirations of the organisation for competitiveness (in terms of market positioning, differentiation, core competencies and capabilities);
- reality will focus on what already exists (processes, people, and technologies); and
- reflection which aims to bridge the gap between reality and the vision.

To realise this strategic vision, it is important to recognise and understand the information and communication needs of various project teams and the challenges associated to linking the various channels for seamless information exchange and effective communication. Design managers have an important part to play here, operating at the boundary of the organisation and interfacing with a wide variety of organisations and collaborators; thus perfectly placed to reflect on the reality and the vision. Case Study 8A gives an insight into the approach that Arup has adopted for strategic project information management.

Case study 8A

Project information management at Arup

Darshan Ruikar and Paul Hill, Arup, UK

The use of technologies in the design, project management, governance, and coordination of large projects raises particular challenges: human (skills, knowledge and managing change), procedural and technical. Also, the modern working environment is more challenging than it has been in the past. These challenges include:

- ever tighter deadlines and shorter project schedules;
- increased requirements for real time collaboration;
- greater data exchange with huge compatibility issues;

- increased use of IT with concomitant growth in data volumes;
- wider geographical distribution of project teams; and
- the 'instant' nature of email and other modern communication methods forcing the speed of decision making.

At Arup it is ensured that we learn from, and build upon, our experiences from past and present projects. We recognise that in a business environment of increasingly large and complex projects, there is a growing need for collaboration and communication within and between organisations.

Internal research at Arup on major projects – M6 Toll, High Speed 1 (formerly CTRL) Olympics 2012 and Rail Link, allowed us to recognise the importance of Information Management (IM) strategies. These included not only software standards, but also process and people standards. As you would expect, Arup's project role and working location has a significant impact on project processes and tools, for instance on the ability to stipulate and freeze software application usage throughout a project's duration (which is only possible if the project team sits outside the Arup network). This was the strategy adopted for the Rail Link Engineering project where the majority of systems were established from 1996 onwards, as well as on the Olympics 2012 when these were decided in 2005. Although this provided system stability, reduced data transfer issues and reduced training costs; it also resulted in the use of out of date systems at the end of the project. This was an important lesson learned for future implementation.

Ensuring that we learn from, and build upon, our experiences from past and present projects, is a constant challenge. The questions we try to address are as follows. What are our experiences and learning about the role of IT in design, project management, coordination and governance on previous infrastructure projects? How are lessons transferred across Arup and onto current infrastructure projects and how can the transfer be improved? Although there are a number of internal knowledge management systems and networks within Arup, previously within Arup IM lessons learned such as in the case of Rail Link, they were reliant on being transferred with people as they moved from project to project. Very little explicit knowledge sharing of project IM, on Arup Projects or within internal skills (knowledge) networks, was evident. Learning from these lessons, it was thought necessary to develop a Project Information Management (PIM) toolkit, which included standards and procedures, captured client requirements and baseline system capabilities, and above all safe guarded against knowledge loss.

Information management

At Arup we believe that Information Management incorporates the management and stewardship of the project's intellectual property. Successful development and implementation of an information management strategy and the associated systems will deliver value through:

- reducing risks – one version of the truth;
- improving efficiency – data reused rather than recreated;
- improving decision support – data available to all who need it.

Information management is a combination of people, process and technology. It is central to a project's success and involves all project stakeholders. Good IM ensures that the project data required is accessible when it is needed, and provided at a cost and quality that meets project team's requirements throughout the project's lifecycle. Moreover, poor quality control over managing project information and lack of document audit trail leads to little confidence that the latest versions of documents and drawings are being used, resulting in greatly increased time spent finding, verifying and reworking information. This leads to inefficiencies within the organisation and also across the entire temporary project organisation.

IM standards and capability

In Arup, IM standards and capability support efficient working and communication between teams and form the foundation and structure for collaboration between project teams. Effective IM standards cost effectively increase project performance from the level of the lowest common denominator to a defined and agreed level.

Whole lifecycle IM strategy

In order for information management to deliver any of the potential benefits within a construction project, the project IM strategy should be developed as part of the overall Project Execution Plan and must be properly implemented. It is also recognised that the value of the information created during the project will increase if a whole life cycle IM strategy is developed and adopted thus aiding the project manager to meet the client's long-term requirements. This strategy will facilitate data reuse and ensure it is built upon, and enriched, as it moves through the subsequent stages of the project. The IM strategy for information exchange covers many areas including:

Processes

- document and email management;
- data management;
- collaboration protocols;
- future use of data i.e. facilities management.

IT systems

- CAD productivity tools;

- mobile technologies;
- geospatial information systems (GIS).

The value of the information created on the project will increase greatly if a whole lifecycle IM strategy is developed and adopted. This strategy will facilitate data reuse and ensure it is built upon and enriched as the project moves through its subsequent phases, right through to completion and beyond. Information created during the project is used for facilities management to support refurbishment and future alteration for many years. The quality of the information can affect the costs of operation, modification and disposal of the installation, as noted by Gallahar et al., 2004) '$15.8 billion in annual interoperability costs were quantified for the capital facilities industry in 2002...two-thirds are borne by owners and operators...predominantly during ongoing facility operation and maintenance (O&M).'

Project IM toolkit

Arup project managers are increasingly asked to advise the client on the information systems required for projects and effective IM strategies have now been implemented on several Arup major projects. The Project IM Toolkit is a standard resource and part of Arup Project Systems, to support medium and large projects. It is intended to provide a head start to project teams and project managers to apply effective and consistent information management, while ensuring that all critical areas are addressed. It builds upon processes and practices that have been successfully used and trialled in Arup projects and develops those into standard approaches. The IM toolkit covers project initiation, application selection, document numbering, folder/metadata structures, processes (document creation, review, approval, distribution, etc.), GI and CAD management, roles and responsibilities and reporting.

The toolkit also draws upon the principles of the Avanti Project Information Management Standard Method and Procedure and on BS1192:2007 collaborative production of architectural, engineering and construction information – code of practice, to ensure our approach is aligned to industry best practice. It can also be used as a foundation for a more consistent approach to information management within Arup.

Conclusion

There are many factors that need consideration when defining strategies for collaborative working. At one end of the spectrum are static and hard issues pertaining to system capability, such as platforms, software and hardware requirements; while at the other end are the more dynamic, soft issues concerning culture, working methods, governance, motivation, competence and

136

absorption (or learning) capacity of end users. Noteworthy management literature often stresses that the capability of an organisation is its demonstrated and potential ability to accomplish against the opposition of circumstance or competition, whatever its sets out to do. In other words, we are concerned with the organisation's ability to develop dynamic capabilities that are responsive to both changing business and to market variables. The term dynamic refers to the capacity to renew competences so as to achieve congruence with the changing business environment. Certain innovative responses are required when time-to-market and timing are critical. The rate of technological changes are rapid therefore determining the nature of the future competition and markets becomes doubly difficult. The term capability entails adaptation, integration, and reconfiguration of internal and external organisational skills resources and functional competences to match the requirements of the rapidly changing environment. The dynamic capabilities approach, thus, seeks to provide a coherent framework which can both, integrate existing conceptual and empirical knowledge, and facilitate prescription. A key step in constructing a conceptual framework related to dynamic capabilities is to identify the foundations upon which distinctive, unique and difficult to replicate advantages can be built, and enhanced.

While the basic tools of strategic analysis to assess an organisation's resources and capabilities remain valid and robust, since the aftermath of the late 1990s technology development, technology can no longer be viewed in isolation. The approach adopted needs to be a holistic, integrated approach that not only takes existing technologies in the context of the business, but also emerging technologies. Also, the multi-disciplinary and collaborative nature of the construction industry necessitates the development of 'collective' competence and multi-party commitment for seamless and successful project delivery. Interoperability between different systems is a major issue due to the existence of multiple teams with varying levels of technology competence.

In collaborative complex projects such as construction projects facilitated by collaborative systems, this will involve software systems operated by different companies and consequently differentially sourced. This can increase the level of difficulty associated with interoperability. Issues such as these, which are common to complex systems can potentially be resolved by adopting innovative, inter-disciplinary approaches such as those promoted by systems engineering. A systems engineering approach challenges the traditional single-discipline approach. This interdisciplinary approach to engineering systems is inherently complex, since the behaviour of and interaction among system components are not always well defined or easily understood. The key consideration in system engineering, besides the system itself, is the human element. From a technology implementation perspective, human response to technology implementation can be a crucial factor, influencing the successes or failure of the implementation.

Factors associated with human issues can be viewed from different perspectives. The introduction of, or alterations to, IT systems have a powerful behavioural and organisational impact. It transforms the way various individuals and groups perform and interact. The implementation of the new systems changes the way information is defined, accessed, and used to manage the resources of the organisation, which lead to new distribution of authority and power between the older and the younger members of staff. These internal organisational changes can breed resistance and opposition, which can often lead to staff reverting to 'old ways'. Still on the same issue, but from a different perspective, staff might fear an increase in workload in the short term, as they bear the burden of responsibility for the successful use of the new package on the initial projects. And as result, they might not be willing to adopt the software. Hence, effort must be directed towards changing staff behaviour and increasing acceptance through proactive measures that engage 'affected' staff in pre-implementation meetings and staff training exercises.

A systems engineering approach examines teams, not individuals. It addresses both technical and business issues. In doing so it focuses on the 'bigger picture'. There lies the big challenge, especially since adopting IT systems is a complex process. At the heart of this complexity is the consideration of a maze of dynamic, interactive, inter-connected and inter-disciplinary activities, systems and components. It is this complexity that invites consideration of strategies for developing new capabilities. Indeed if management of scant resources is the basis of financial profits, then it follows that issues such as skills acquisition, management of knowledge, know-how, and learning (knowing why and why not) become fundamental strategic issues. It is in this 'human' dimension, encompassing cooperation, collaboration, skills acquisition, retention, learning, and accumulation of valuable intangible knowledge assets where the greatest potential for contributions to strategy lies.

Further reading

For further information on technology strategies see *ITcon Special Issue on Strategies for Collaborative Working*. Ruikar, K and Emmitt, S (2009) Editorial – Technology Strategies for Collaborative Working, ITcon, Special Issue Technology Strategies for Collaborative Working, 14, pp.14–16, Full text: http://www.itcon.org/

9

FUTURE DIRECTIONS

A number of recurrent themes are brought together in this final chapter to try and highlight some of opportunities and challenges associated with design management in a collaborative environment. Some of the case study contributors have already provided their views on the future role of the construction design manager. They have, independently, emphasised the role of technologies (especially BIM) as an agent of change and also highlighted the need for better understanding of the role within the construction sector. Our approach in this chapter is to a take a reflective stance and set out some of the more important issues facing design management within the AEC sector. In doing this, the one word that constantly surfaces is 'understanding'. Throughout the book we have been dealing with understanding, specifically understanding the role of the design manager and by implication, a better understanding of the complexities of design.

Better understanding of design

Contractors have always been good at the business aspects of construction, but have not always had a sufficient appreciation of design; or more specifically the challenges associated with generating and managing design. From the small body of published research and in our conversations with contractors and construction design managers it is clear that this has started to change. Contractors have become much more aware and sensitive to the value that design can add to their businesses and customers. This increase in understanding has gone hand in hand with the development of the construction design management role. The better understanding of design has also resulted in contractors giving greater attention to the management of the early design stages. It is here that design risks can be designed out and design value designed into the project. More specifically, the importance of client briefing in clearly establishing what is required, the design parameters, before starting to design (or build) is much better appreciated compared to a decade ago. Similarly, the importance of design development and detailing the design in ensuring problems are designed out before work starts on site.

There are, however, some differences between new build and refurbishment projects which relate to risk and uncertainty. In new build projects there may be some degree of risk dealing with unknowns, such as demolition of existing buildings and the risk of damage to adjacent properties. And, no matter how thorough the site investigation there is always an element of risk until the building is out of the ground. But once out of the ground there should be very little uncertainty or risk to the project. This presumes that the pre-construction activities of briefing, design and detailing have been carried out effectively and that the design managers have already resolved clashes and coordinated the information. In contrast to new build schemes, refurbishment projects tend to carry increased risks in terms of what is unknown. It is only when the building fabric is opened up that the risks can be fully understood. Thus the construction design manager's role tends to be more concerned with risk and dealing with uncertainties that affect the development of the design. This calls for different skills, and one might argue different levels of sensitivity to historical building fabric.

Better commercial awareness

It is becoming popular for contracting organisations to offer a 'one stop shop' or 'total solution' for their customers. This reflects a desire for integration and collaboration as well as a desire to expand the services provided and hence improve the business opportunities for the contractor. Pugh's (1990) work on total design is in many ways the precursor of the total solution provider. Total design encompasses product, process, people, and organisation, covering the entire lifecycle of a product, from recognition of a market need through to the satisfaction of that need. At the core of total design is design value. Relating Pugh's work to construction it is entirely feasible that contracting organisations can provide a total (integrated) service to their clients. What is uncertain from the literature is how the construction design manager's role is affected by total solution provision and vice versa. However, one theme that constantly emerges when we speak to contractors is the need for (construction) design managers to be better equipped with an understanding of the commercial drivers behind contracting organisations and projects. This appears to be an area that could be better addressed in the education of undergraduate and trainee design coordinators and design managers.

Better understanding of people

We are unique as individuals and also unique in the time and space we occupy. Designers, in the widest sense of the word, are also unique in their areas of specialism, be it for example in the design of a structural frame, services, interiors or the façade of a building. Collaborative working makes use of diversity and harnesses individual skills and knowledge to create equally unique buildings – this means that is crucial we understand the people with whom we collaborate. Understanding what others can contribute, and equally what they cannot, can tap into a collective creativity, which may otherwise be ignored and therefore the potential of our 'uniqueness' not fully exploited.

There is a very old saying that 'a fool with a tool should be given a wide berth' and by extension 'a fool with a foolish tool is dangerous in a business setting'. By this we mean that it is important to understand people and allocate the most appropriate person to a specific task or role. Putting a person in the wrong position or role will not only induce a high degree of (unnecessary) stress for the individual, but will compromise performance. In terms of tools, it also follows that we need to apply the most appropriate managerial tools and techniques (and supporting technologies) to business organisations and the projects that fuel them. Getting this wrong will have a negative impact on those using and affected by them.

Understanding and engaging clients

There is no doubt that clients (building sponsors) have become more sophisticated in their demands of those designing and realising their buildings. Clients have also become more demanding in terms of the quality, performance and functionality of their buildings. They have started to be more demanding when it comes to technologies, with many clients demanding that their buildings are designed within a single virtual model, partly to help with efficiencies in design and realisation and partly so that they retain a virtual model at the end of the project. This will then be used to help with the management of the facility over its service life.

Understanding and engaging building users

There is not a particularly strong tradition of user involvement in the design process in the UK, compared to, for example, the Scandinavian countries. However, efforts continue to be made to incorporate more effectively user values within the initial briefing and design development phases. Again, this is an area that is ripe for exploration and development; and one that may be ideally suited to the design manager role.

A better understanding of construction design managers

We have been aware for some time that there are many people working within the construction sector who do not understand what a construction design manager does. This spans from individuals in charge of organisational training schemes through to other professionals and trades. It has been, and continues to be a major concern of Loughborough University's AEDM graduates. However, as the role develops so too will other participants' understanding of the value design managers add to construction projects.

Better understanding of technologies

Of course the usual drivers (economic, environmental, legislative, social, technological, etc.) are constantly at play and so it is not necessary to discuss these in detail, merely to recognise that they will continue to influence future developments in the field. Building in times of austerity tends to focus attention on efficiency and effectiveness in all aspects of a business, and this is not such a bad thing if it is done considerately and sensitively.

As discussed throughout this book, the role of information technologies as an enabler of collaborative design management is considerable. Of all the factors that are likely to shape the role of the construction design manager, building information modelling has to be the most significant. Given that the greatest value of managing design is in the early stages, it is inevitable that the design management role will move from the site to the (virtual) design office, and with it the design manager's role will evolve further. Here the design manager will be able to make a positive contribution to the development of the project, helping to build in design value and eradicating design waste in the virtual model. As BIM technologies mature and are taken up more widely, there should be less need for design managers at the post-contract phase, signifying a further shift in the position (and importance) of the design manger role.

Functionalities of evolving collaborative systems (e.g. BIM tools) extend beyond traditional organisational boundaries to accommodate the collective activities within the project. Developments in computer-mediated collaborative working environments have spearheaded a shift from organisation-focused approaches to collaborative, lifecycle and project-centred approaches. Intrinsic to this is a shift in the nature of the knowledge that is contained within: knowledge that resides in the organisation and that which resides in the collaborative project setting. This lifecycle knowledge, links the 'harder issues' concerning the system capability (i.e. platform, software and hardware) with the 'softer issues' concerning the human component of habits of working, attitudes, motivation, competence and absorption capacity. This change to lifecycle knowledge management requires considerable effort towards integrating the various disparate systems and processes that reside within the project. In other words, it calls for a change management strategy that recognises the inevitability of the change and plans for it.

Important aspects to consider when planning for such change are recognising that a computer mediated collaborative work setting is a complex social process, and therefore change may be met with resistance. Often the resistance stems from deep seated concerns of the unknown concerning changes to roles, responsibilities, and processes, fear of redundancies, possible increases to workloads, anxiety about using new tools, and generally concerns that stem from stepping into the unknown. Central to a successful change management strategy is recognising that resistance is often rational and justified. Understanding the reasons for resistance and addressing potential objections, before implementation begins will go along way towards facilitating change. Adequate and high profile multi-dimensional requirements capture involving all stakeholders (internal and external) will go a long way towards achieving this. So too will the involvement of the senior design managers in encouraging and stimulating an appropriate attitude towards change. It is therefore expected, that through a multi-dimensional approach requiring the effective leadership of the process, people and culture, a greater adaptability and acceptance of change can be created.

Case study 9A

Collaborative design management – still evolving

Paul Wilkinson, pwcom.co.uk Ltd

The development of the world wide web during the 1990s started a technology change across architecture, engineering and construction and facilities management that continues to this day. For an often geographically-dispersed, multi-company project delivery team, the ability to send messages and associated attachments electronically was attractive, but email was not the magic bullet that design teams might have anticipated. Paper-based drawings and documents still dominated, and most of the processes involved in their exchange were simply replicated electronically. It was difficult to ensure everyone worked from most up-to-date version of a drawing. Email 'conversations' could not easily be tracked across different companies' email systems; and people could be swamped with messages needlessly copied to them 'for information'.

In the late 1990s/early 2000s, web-based construction collaboration applications – 'extranets' – emerged as a potential alternative to email and paper-based communications. Instead of each company maintaining its own island of information, a single secure shared repository of project data was created, hosted by specialist Software-as-a-Service (SaaS) providers. So long as they had internet access and a web-browser, project team members could access the latest project information, annotate drawings and documents, and manage common construction processes (requests for information, change orders, etc.) online. An electronic audit trail recorded who did what and when, creating more transparent reporting than was ever possible via email.

The 1994 Latham and 1998 Egan reports also encouraged more collaborative approaches to construction project delivery. In 2002, *Accelerating Change* identified information technology as a cross-cutting industry issue, and recommended wider adoption of electronic systems to manage design and construction information. All very well, but successful collaboration was – and remains – more about people and processes than technology (vendors often talked about an 80:20 split).

Email, however, is still important, particularly for minor projects undertaken by small teams and involving limited exchange of design information. Other document management systems are also used. These ranged from simple FTP (file transfer protocol) sites, through intranet-type applications such as Microsoft SharePoint, to construction-specific tools that can be hosted by construction businesses.

Building information modelling, however, poses new challenges. Instead of exchanging various deliverables – written briefs, specifications, bills of quantities, design drawings, visualisations, photographs, contracts, forms, notices, etc. – project teams will be developing and progressively populating accurate digital models of the built asset. Design collaboration will increasingly be about structured data – not drawings or documents. Models will incorporate large amounts of information, in addition to the dimensions and geometry of the various components and materials comprising the asset, and some data will be hyper-linked to yet further information held in other data sources.

Implicit in the BIM approach is an assumption of collaboration, with designers sharing data with the ultimate client, with contractors, component manufacturers and materials suppliers. This shared, jointly developed data will be produced to higher degrees of accuracy, and poses new questions about contracts, data ownership, liability, and commercially sensitivity. And as these issues are resolved, existing project roles and responsibilities are likely to evolve still further.

In some respects, the construction sector is following a path already beaten by the aerospace and automotive industries. Computer driven design and manufacture created more certainty about what was required and when. Processes became progressively leaner, with greater precision, less waste, shorter supply chains and more just-in-time delivery.

Off-site fabrication and modularisation demonstrate that construction is already changing, and some BIM commentators see this trend accelerating. Ray Crotty (2011), for example, has predicted that seamless BIM data 'will help unify the industry's supply chains, freeing construction from its craft origins, transforming it into a modern, sophisticated branch of the manufacturing industry.'

For the owner or operator of a built asset, this BIM enabled, streamlined production process should see design and construction data flow seamlessly into information systems used for operation, maintenance and facilities management. Here it could also be connected to real time data sources (from environmental indicators such as power consumption, temperature, humidity, light intensity, etc., to business productivity or other indicators showing how fit for purpose the asset has proved). Indeed, post-occupancy evaluation by end users could become a key building block for design collaboration on future schemes.

Education and training

Given the rapid changes surrounding the development of the construction design manager role, a number of questions relate to the education and training of design managers.

University education

As mentioned in the Introduction there are currently very few educational programmes dedicated to design management. The programme at Loughborough University has matured since its launch, evidenced by the changes made to its content. Although the programme places much emphasis of sustainable (low carbon) design and interdisciplinary working, the programme has become much more focused, helping to set it apart from construction management or architectural technology programmes. Generic management and project management modules have been replaced with modules that are tailored to the construction design manager, such as 'Principles of design management' in the first year and 'Collaborative design management' in the second year. These modules contain a mix of theory and practical skills, providing the student with the knowledge and ability to apply in industry. The drawing and information communication technology modules have also evolved, with emphasis moving from 2D and 3D drawing to virtual modelling and, more recently, increased attention to collaborative technologies. This evolution has been achieved through collaboration of academics, students and industrialists. The goal is simple, to be the best at what we do; and we can only do that via effective collaboration with our students and industrial partners.

The philosophy at Loughborough University has always been to keep the intake of the AEDM programme low and emphasise quality rather than quantity. Although this approach has been highly effective for our students and their employers, in taking this approach we are conscious that there are many potential applicants who have to go elsewhere. Given the year-on-year increase in demand for our programme from students, and demand for our graduates from industry, we would expect to see other universities starting to offer very similar courses. We feel that the construction sector needs both graduates who are educated in design management and also experienced professionals who have moved into a design management role, combining academic and experiential learning to the benefit of the project and the organisation. And here the professional bodies can help by accrediting design management courses (as currently undertaken by the CIOB) and promoting the role (again, as currently being done very effectively by the CIOB).

Training and development

The majority of training for design managers is conducted in house by most of the large contracting organisations, with the tendency reducing with size. The CIOB's recent survey of design management training found that employers needed to take their responsibilities more seriously when it came to providing clearly defined training or mentoring schemes (Keilthy, 2011). Our case study contributors have helped to demonstrate some good practice in this regard, with training schemes addressing the inter-related aspects of:

- Task-based skills. These are typically concerned with legal and contractual arrangements, approvals, document management systems, change management procedures, technical skills, etc., and are tailored to the employer's existing processes and procedures.
- Socio-emotional skills. These are typically concerned with how to work with others and tend to address issues such as negotiation, communication and leadership skills, and are specific to the employee.

The role of the CIOB

In addition to accrediting design management programmes, the CIOB has also been very active in promoting design management to its members, especially through articles in its journal *Construction Research and Innovation (CSI)*, (see for example; Keilthy, 2011; Eynon, 2011). The CIOB's Design Management Group has been instrumental in producing a book, the *Design Manager's Handbook* (CIOB, 2013), which provides guidance on the role and also includes a design management code of practice.

Case study 9B

The future of construction design managers

*John Eynon, Chair of CIOB's Design Management Group/
Open Water Consulting*

Over the last 12 months I have been working on the CIOB Design Manager's Handbook as part of an ongoing project on Design Management at the Chartered Institute of Building (CIOB). Over the same period, interest in BIM has rapidly increased, pushed both by the UK government construction strategy led by Paul Morrell (the Chief Construction Advisor), Mark Bew and David Philp, and also by the innovators and early adopters in the construction industry such as Laing O'Rourke, Balfour Beatty, VINCI, BAM and others who have continued to blaze the BIM trail.

As my own design management/BIM journey has continued, I have begun to wonder whether these paths are converging. BIM is not design management, but it enables design management to happen on a level previously unattainable and beyond the wildest dreams of those who have worked at the coal face. Conversely design management isn't BIM either, but if BIM is the management of data over the whole built asset lifecycle at every stage, isn't linking every stakeholder with the information they need, design management in its purest sense? When everything is reduced to the absolute minimum, is not design management about information and data management?

Design management is about connection and flow: ensuring that all aspects of the project team are connected, that communication is happening as it should, and that information of the right kind is flowing to the right people at the right time. It is also concerned with making sure that the right people are talking to each other at the right time, for example specialist supply chain and designers are having the 'right' conversations. In a way, design management is the glue that holds this whole process together.

In the BIM world there are still activities and workflows, but some of the dynamics change. Those that once sat back and waited for information to hit them can now get it for themselves by accessing the virtual model. Different cultures and different mindsets are now required in order to work effectively in a BIM environment; and what of the design manager in a BIM environment? For a relative newcomer to the construction discipline all change (again) seems to be order of the day.

Already there is a ground swell of opinions about the need for a 'model manager', 'integrator', 'BIM manager', 'coordinator', 'data manager' – someone who sits at the centre of this process and ensures that it all happens as it should. This person, or perhaps a team of people on some projects, will need to be sufficiently BIM savvy to know what needs to happen but also sufficiently knowledgeable about the design and construction process in the relevant sector to be able to translate this into the reality of collaborative teamworking; i.e. it calls for an understanding of the roles, people, businesses and organisations involved.

In the UK currently we are 'crossing the chasm' to use Geoffrey Moore's terminology. This is the chasm between innovation and early adoption, into mainstream industry use. Up to now, BIM development has been predominantly driven by the software houses and techno geeks. However, for BIM to become accepted industry practice, then it is the hardnosed CEOs and middle managers that need to be convinced. They're mildly entertained by new widgets and gizmos but really they want to know about investment costs, payback periods, profitability and return on investment. They need a different spin on the BIM story, different messages and ideas from the ones that got us here.

This is a difficult period as we struggle with the economic situation and an industry in the throes of evolution and industrialisation, but those that have jumped into BIM early are already seeing the benefits through better collaborative working – less waste and more efficiency – achieving more with less. In time we won't mention BIM at all – we'll just do it, no big deal. It's rather like when CAD became mainstream, but only more so. When everyone is working this way, and all parties involved know how to work in the 'common data environment' of the BIM world, there will still be people coordinating, marshalling, and managing data and information over all phases of a project. It may not be one, or the same person, over the whole process. Already we can see that at different stages leadership of the model may be in different hands, for example the client, designer or contractor. The only difference now is that we have a mechanism that prevents loss of knowledge as we transfer leadership and ownership.

To me, this is the next step for design managers, for those that want to make this evolutionary step. Design managers know the process and they understand the lifecycle and the inputs and personalities involved. The best construction design managers naturally and intuitively integrate, collaborate and communicate, building relationships, drawing people and information together, resolving conflicts, enabling things to happen as they should, and improving how things work along the way. BIM is the perfect environment in which to make this happen; indeed Latham and Egan *et al.* dreamt of this day and BIM could help make it an everyday, every project, reality.

Some construction design managers will gravitate to the technology end of the spectrum (geek world) and others will remain firmly at the process end, but they will know enough of the technology to have a handle on it and to lead and manage the process. This is not for the fainthearted – change never is, but I firmly believe that design managers are ideally placed to move into the world of BIM and own it – leading, coordinating, integrating, innovating and creating real value for customers and everyone involved – truly kings of the building lifecycle or whichever stage one wishes to identify with. Now let me see . . . where do I sign up for that BIM course?

Research

Compared to many avenues of inquiry, such as project management and construction management, the design management field has not been researched as extensively. One of the reasons for this is that it is a relatively new role and an emerging discipline. In such cases, it is common for researchers to spend time trying to make sense of what is happening in the sector. Given the rapid establishment of the discipline and the positive mutation of the role, it is not an easy task for researchers to keep up with developments. The peer reviewed journal *Architectural Engineering and Design Management* has been leading the field in publishing research about design management and dedicating a number of special issues to aspects of design management (see Emmitt, 2007, 2009, 2011). A small number of articles have also been published in other journals; however, collectively this body of knowledge is small compared to related fields such as construction project management. Opportunities for undertaking research are extensive, and the potential for providing some unique insights and contributions could come, for example, from exploring the:

- relationship between design managers and other collaborators (e.g. facilities managers);
- changing role of design managers with the uptake of BIM and associated collaborative technologies;
- differences in approaches to design management in new build and refurbishment projects;
- relationship between projects and business and the role that effective design management plays;
- interpretations and applications of design management in different international settings;
- effect of collaborative procurement on design managers' roles and responsibilities.

Final words

There has been some debate as to whether the construction design management role is a discipline in its own right. In a collaborative environment, it is crucial that skills and knowledge are given preference over titles, however, from an individual's and organisation's perspective it is important for people to have meaningful job titles. While many individuals will undertake some aspects of design management in their daily tasks, it is clear that there is a specific role (or roles) for the design manager. And if this role continues to add value to the process and the building, then the role is to be embraced and nourished.

Given the rapidly evolving nature of the construction design manager role we are aware that we can only deal with rather generic issues within this book. It is important that individuals and organisations take the ideas and apply them in an appropriate manner to their own unique circumstances. This is particularly true of those working in countries other than the UK, where contextual issues such as culture and legislation have a major bearing on how design management is applied in industry. In this book, our focus has deliberately been on the UK; however we are aware of the major developments in the field of design management in countries such as Brazil, China and the USA.

Further reading

One of the few peer-reviewed journals dedicated to the field of design management is *Architectural Engineering and Design Management*. This journal has dedicated a number of special editions to design management and continues to publish articles reporting various aspects of design management. For insights into generic design management research, see *Design Management: exploring fieldwork and applications*, edited by Jerrard and Hands (2008, Routledge).

REFERENCES

Affleck, A. (2009) 'Take control of podcasting on the Mac', version 2.1, TidBITS Publishing Inc.

Allinson, K. (1993) *The Wild Card of Design: a perspective on architecture in a project management environment.* Oxford: Butterworth-Heinemann.

Allinson, K. (1997) *Getting There by Design: an architect's guide to design and project management.* Oxford: Architectural Press.

Archer, B. (1967) 'Design management', *Management Decision*, 1(4): 47–51.

Arnold, D.N. and Rogness, J. (2008) 'Moebius transformations revealed'. Available at http://www.ima.umn.edu/~arnold/ (accessed October 2008).

Austin, S., Baldwin, A., Hammond, J., Murray, M., Root, D., Thompson, D. and Thorpe, A. (2001) *Design Chains: a handbook for integrated collaborative design.* Tonbridge: Thomas Telford.

Baden Hellard, R. (1995) *Project Partnering: principle and practice.* London: Thomas Telford.

Basu, S. and Deshpande, P. (2004) 'Wipro's people processes: a framework for Excellence', White Paper. Available at: http://www.wipro.com/insights/wipropeopleprocesses.htm (accessed 23 April 2004).

Beier, K.P. (2008) 'Virtual reality: a short introduction'. Available at: www-vrl.umich.edu/intro/ (accessed 15 June 2010).

Beim, A. and Jensen, K.V. (2007) 'Forming core elements for strategic design management: how to define and direct architectural value in an industrialized context', in Emmitt, S. (ed.) (2007) 'Aspects of design management', *Architectural Engineering and Design Management*, special edn., 3(1): 29–38.

Bennett, J. and Jayes, S. (1995) *Trusting the Team: the best practice guide to partnering in construction* Tonbridge: Thomas Telford.

Best, K. (2006) *Design Management: managing design strategy, process and implementation*, AVA, Lausanne.

Best, K. (2010) *Fundamentals of Design Management.* Lausanne: AVA.

Bibby, L. (2003) 'Improving design management techniques in construction', EngD thesis, Loughborough University.

Boddy, D., Cahill, C., Charles, M., Fraser-Kraus, H. and Macbeth, D. (1998) 'Success and failure in implementing supply chain partnering: an empirical study', *European Journal of Purchasing and Supply Management*, 4, 143–51.

Borja de Mozota, B. (2003) *Design Management: using design to build brand value and corporate innovation.* New York: Allworth Press.

Boyle, G. (2003) *Design Project Management.* Aldershot: Ashgate.

Brawne, M. (1992) *From Idea to Building: issues in architecture.* Oxford: Butterworth Architecture.

Brunton, J., Baden Hellard, R. and Boobyer, E.H. (1964) *Management Applied to Architectural Practice*, George Goodwin for The Builder, Aldwych.

BuildingSMART (2012) *Glossary*. Available at: http://www.buildingsmart.com/content/glossary_terms (accessed 13 June 2012).

Cardell, S. (2002) *Strategic Collaboration*. London: Hodder Arnold.

Churchman, C.W. (1967) 'Wicked problems', *Management Science*, 4(14): 141–42.

CIOB (2013) *Design Manager's Handbook*, CIOB/Wiley-Blackwell, Chichester.

Cooper, R. and Press, M. (1995) *The Design Agenda: a guide to successful design management*. Chichester: Wiley.

Cox, A. (2001) 'Understanding buyer and supplier power: a framework for procurement and supply competence', *The Journal of Supply Chain Management*, 37(2): 8–15.

Cross, N. (2011) *Design Thinking: understanding how designers think and work*. Oxford: Berg.

Crotty, R. (2011) *The Impact of Building Information Modelling: transforming construction*. Abingdon: Spon Press.

Cullen, J.B., Johnson, J.L. and Sakano, T. (2000) 'Success through commitment and trust: the soft side of strategic alliance management', *Journal of World Business*, 35(3): 223–40.

Dalziel, B. and Ostime, N. (2008) *Architect's Job Book*, 8th edn. London: RIBA Enterprises.

Das, T. and Teng, B. (2000), 'A resource-based theory of strategic alliances', *Journal of Management*, (26): 31–61

Daugherty, P., Glenn Richey, R., Roath, A.S., Min, S., Chen, H., Arndt, A.D. and Genchev, S.E. (2006), 'Is collaboration paying off for firms?' *Business Horizons*, 49: 61–70.

DCLG (2007) *Building a Greener Future*, policy statement, Department of Communities and Local Government. London: The Stationery Office.

Design Management Institute, Boston, MA (www.dmi.org)

Eastman, C.M., Teicholz, P., Sachs, R. and Liston, K. (2011), *BIM handbook: a guide to building information modelling for owners, managers, designers, engineers, and contractors*, 2nd edn. Hoboken, NJ: John Wiley & Sons.

Egan, J. (1998) *Rethinking Construction, Report of the Construction Task Force on the Scope for Improving the Quality and Efficiency of the UK Construction Industry*, Department of Environment, Transport and Regions (DETR). London: HMSO.

Egan, J. (2002) *Rethinking Construction: accelerating change*, Strategic Forum for Construction. London.

Emmerson Report (1962) *Survey of the Problems before the Construction Industries*. London: HMSO.

Emmett, B. (2002) 'IT service management: people + process + technology = business value', *The IT Journal*, third quarter 2002. Available at: www.hp.com/execcomm/itjournal/third_qtr_02/article2a.html (accessed 6 April 2004).

Emmitt, S. (1997) 'The diffusion of innovations in the building industry', PhD thesis. University of Manchester.

Emmitt, S. (1999a) *Architectural Management in Practice: a competitive approach*. Harlow: Longman.

Emmitt, S. (1999b) 'Architectural management: an evolving field', *Engineering, Construction and Architectural Management*, 6(2): 188–96.

Emmitt, S. (2007a) *Design Management for Architects*, Oxford: Blackwell.

Emmitt, S. (ed.) (2007b) 'Aspects of design management', *Architectural Engineering and Design Management*, special edn, 3(1).

Emmitt, S. (ed.) (2009) 'Design management for sustainability', *Architectural Engineering and Design Management,* special edn, 5(1 and 2).

Emmitt, S. (2010) *Managing Interdisciplinary Projects: a primer for architecture, engineering and construction*. Abingdon: Spon Press.

Emmitt, S. (ed.) (2011) 'Lean design management', *Architectural Engineering and Design Management*, special edn, 7(2).

Emmitt, S. and Gorse, C.A. (2007) *Communication in Construction Teams*. Abingdon: Spon Press.

Emmitt, S., Prins, M. and Otter, A. (eds) (2009) *Architectural Management: international research and practice*. Chichester: John Wiley & Sons.

Emmitt, S., Pasquire, C. and Mertia, B. (2012) 'Is good enough "making do"?: an investigation of inappropriate processing in a small design and build company', *Construction Innovation*, 12(3): 369–83.

Eynon, J. (2011) 'Design management: a role evaluation', *Construction Research and Innovation*, 2(2): 42–6.

Farr, M. (1966) *Design Management*. London: Hodder and Stoughton.

Forsyth, D.R. (2006) *Group Dynamics*, 4th edn. Belmont, CA: Thomson Wadsworth.

Frankel, R., Smitz Whipple, J. and Frayer, D.J. (1996), 'Formal versus informal contracts: achieving alliance success', *International Journal of Physical Distribution and Logistics Management*, 26(3): 47–63.

Fuji Xerox (2003) 'Aligning people processes and technology'. Available at: www.fujixerox.com.au (accessed 23 April 2004).

Gallaher, M.P., O'Connor, A.C., Dettbarn, J.L. Jr. and Gilday, L.T. (2004) 'Cost analysis of inadequate interoperability in the US capital facilities industry', NIST GCR 04-867, pp. 194, August. Available at: http://fire.nist.gov/bfrlpubs/build04/PDF/b04022.pdf.

Gohil, U. (2010) 'Value enhanced collaborative working (VECW)', EngD thesis, Loughborough University.

Goolsby, C. (2001) 'Integrated people + processes + tools = best-of-breed service delivery', Getronics White Paper. Available at: http://itpapers.news.com/ (accessed 23 April 2004).

Gorse, C.A. (2002) 'Effective interpersonal communication and group interaction during construction management and design team meetings', PhD thesis, University of Leicester.

Gray, C. and Hughes, W. (2001) *Building Design Management*. Oxford: Butterworth-Heinemann.

Gray, C., Hughes, W. and Bennett, J. (1994) *The Successful Management of Design*. Reading: Centre for Strategic Studies in Construction.

Green, R. (1995) *The Architect's Guide to Running a Job*, 5th edn. Oxford: Butterworth Architecture.

Grieves, J. (2000) 'Introduction: the origins of organizational development', *Journal of Management Development*, 19(5), 345–447.

Grilo, L., Melhado, S., Silva, S.A.R., Edwards, P. and Hardcastle, C. (2007) 'International building design management and project performance: case study in São Paulo, Brazil' in Emmitt, S. (ed.) (2007) 'Aspects of design management', *Architectural Engineering and Design Management*, special edn., 3(1): 5–16.

Gropius, W. and Harkness, S.P. (eds) (1966) *The Architect's Collaborative, 1945–1965*, London: Tiranti.

Hart, O. and Moore, J. (1999) 'Foundations of incomplete contracts', *Review on Economic Studies*, 66: 115–38.

Hartley, P. (1997) *Group Communication*. London: Routledge.

Henderson, J.R. and Ruikar, K. (eds) (2010) 'Technology implementation strategies for construction organisations', *Engineering, Construction and Architectural Management*, 17(3): 309–27.

Herkenhoff, L.M. (2006) 'Podcasting and VODcasting as supplementary tools in management training and learning'. Available at: www.iamb.net/CD/CD06/MS/71_Herkenhoff.pdf (accessed 1 July 2012).

Hicks, B.J., Culley, S.J. Allen, R.D. and Mullineux, G. (2002) 'A framework for the requirements of capturing, storing and reusing information and knowledge in engineering design', *International Journal of Information Management*, 22: 263–80.

Huq, Z. (2005) 'Managing change: a barrier to TQM implementation in service'.

IBM (1999) 'Arriving at the upside of uptime: how people processes and technology work together to build high availability computing solutions for e-business', White Paper. Available at: www.dmreview.com/whitepaper/ebizc.pdf (accessed 24 April 2004).

ILO (2001) 'The construction industry in the twenty-first century: its image, employment prospects and skill requirements', International Labour Office, Geneva, industries', *Managing Service Quality*, 15(5): 452–70.

Jerrard, R. and Hands, D. (eds) (2008) *Design Management: exploring fieldwork and applications*. Abingdon: Routledge.

Jones, J.C. (1992) *Design Methods*, 2nd edn. New York: Van Nostrand Reinhold.

Jørgensen, B. and Emmitt, S. (2008) 'Lost in transition – the transfer of lean manufacturing to construction', *Engineering Construction and Architectural Management*, 15(4): 383–98.

Jørgensen, B. and Emmitt, S. (2009) 'Investigating the integration of design and construction from a "lean" perspective', *Journal of Construction Innovation*, 9(2): 225–40.

Kajewski, M.A. (2007) 'Emerging technologies changing our service delivery models. State Library of Queensland, South Brisbane, Australia', *The Electronic Library*, 25(4).

Kalajdzievski, S. (2008) *Math and Art, An Introduction to Visual Mathematics*. Boca Raton, FL.: CRC Press, Taylor & Francis Group.

Keilthy, B. (2011) 'Untapped potential', *Construction Research and Innovation*, 2(1): 46–50.

Kelly, J., Male, S. and Graham, D. (2004) *Value Management of Construction Projects*. Oxford: Blackwell.

Kern, H., Johnson, R., Galup, S. and Horgan, D. (1998) *Building the New Enterprise: people, processes and technology*. Palo Alto, CA: Prentice Hall PTR.

Khemlani, L. (2008) 'AECbytes product review – Navisworks 2012'. Available at: www.aecbytes.com/review/2008/NavisWorks2009.html (accessed 30 May 2012).

Koskela, L. (2004) 'Making do – the eighth category of waste', *Proceedings of the 12th International Group for Lean Construction Conference*. Denmark: Elsinore.

Larkin, B. (2003) 'Aligning people process and technology'. Available at: www.paperlesspay.org/articles/Technology.pdf (accessed 23 April 2004).

Latham, M. (1993) *Trust and Money: interim report of the joint Government industry review of procurement and contractual arrangements in the United Kingdom construction industry*. London: HMSO.

Latham, M. (1994) *Constructing the Team*. London: HMSO.

Laudon, K.C. and Laudon, J.P. (2002) *Management Information Systems*, 7th edn. Englewood Cliffs, NJ: Prentice-Hall.

Lipman, R.R. and Reed, K.A. (2003) 'Visualization of structural Steel product models', *ITcon*, 8, 51–64.

Liston, K., Martin, F. and Winograd, M. (2001) 'Focused sharing of information for multidisciplinary decision making by project teams', *ITcon*, 6: 69–82.

Littlefield, D. (2005) *The Architect's Guide to Running a Practice*. Oxford: Architectural Press.

Lottaz, C., Rud, S. and Smith, I. (2000) 'Increasing understanding during collaboration through advanced representations', *ITcon*, 5: 1–24.

Lu, S. C-Y., Elmaraghy, W., Schuh, G. and Wilhelm, R. (2007) 'A scientific foundation of collaborative engineering', *Annals of the CIRP*, 56: 605–34.

Mackinder, M. and Marvin, H. (1982) *Design Decision Making in Architectural Practice*, Research Paper 19. York: University of York Institute of Advanced Architectural Studies.

Major, E.J. and Cordey-Hayes, M. (2000) 'Engaging the business support network to give SMEs the benefit of Foresight', *Technovation*, 20: 589–602.

Maleyeff, J. (2006) 'Exploration of internal service systems using lean principles', *Management Decision*, 44(5): 674–89.

Mann, D. (2010) *Creating a Lean Culture: tools to sustain lean conversions*, 2nd edn. New York: Productivity Press.

Maskin, E. and Tirole, J. (1999) 'Unforeseen contingencies and incomplete contracts', *Review on Economic Studies*, 66: 83–114

Meng, P. (2005) 'Podcasting and vodcasting: a White Paper – definitions, discussion and implications', University of Missouri: IAT Services. Available at: www.wssa.net/WSSA/SocietyInfo/ProfessionalDev/Podcasting/Missouri_Podcasting_White_Paper.pdf (accessed 10 July 2012).

Mills, F.T. and Glass, J. (2009) 'The construction design manager's role in delivering sustainable buildings', in Emmitt, S. (ed.) "Design Management for Sustainability"', *Architectural Engineering and Design Management*, 5(1 and 2): 75–90.

Mintzberg, H. (1973) *The Nature of Managerial Work*. New York: Harper and Row.

Mudambi, R. and Helper, S. (1998) 'The "close but adversarial" model of supplier relations in the US auto industry', *Strategic Management Journal*, 19: 775–92.

Mulder, I. (2004) 'Understanding designers designing for understanding', PhD thesis, Telematica Institute, University of Enschede, the Netherlands.

Nicholson, M.P. (ed.) (1992) *Architectural Management*. London: E&FN Spon.

Nicholson, M.P. (1995) 'Architectural management: from Higgin to Latham', PhD thesis, University of Nottingham.

Oakley, M. (1984) *Managing Product Design*. Chichester: John Wiley & Sons.

Obonyo, E.A., Anumba, C.J., Thorpe A. and Parkes, B. (2001) 'Agent-based support for electronic procurement in construction', *Proceedings of 8th International Workshop of the European Group for Structural Engineering Applications of Artificial Intelligence* (EG-SEA-AI), CICE, Loughborough University, UK, 20–22 July, pp. 268–79.

OGC (2003) 'Achieving excellence in construction, building on success', London: OGC. Available at: www.ogc.gov.uk/documents/BuildingOnSuccess.pdf (accessed 25 November 2010).

Ohno, T. (1978) (1988 English translation) *Toyota Production System: beyond large scale production*. Portland, OR: Productivity.

Otter, A.F.H.J. (2005) 'Design team communication using a project website', PhD thesis, Bouwstenen 98, Technische Universiteit Eindhoven.

Otter, den A. and Emmitt, S. (2007) 'Exploring effectiveness of team communication: balancing synchronous and asynchronous communication in design teams', *Engineering, Construction and Architectural Management*, 14(5): 408–419.

Pang, D. (2010) 'Barriers, enablers and potential of incorporating cost estimation into a building information model, AEDM final year research project', School of Civil and Building Engineering, Loughborough University.

Phillips Report (1950) *Report of a Working Party to the Minister of Works*. London: HMSO.

Pugh, S. (1990) *Total Design: integrated methods for successful product engineering*. Harlow: Addison-Wesley.

Rekola, M., Makelainen, T. and Hakkinen, T. (2012) 'The role of design management in the sustainable building process', *Architectural Engineering and Design Management*, 8(2): 78–89.

RIBA (1962) *The Architect and His Practice*. London: Royal Institute of British Architects.

RIBA (2007) *RIBA Plan of Work*. London: RIBA Publications.

Ritter, T. and Gemeunden, H.G. (2004) 'The impact of a company's business strategy on its technological competence, network competence and innovation success', *Journal of Business Research*, 57: 548–56.

Rogers, E.M. and Kincaid, D.L. (1981) *Communication Networks: toward a new paradigm for research*. New York: The Free Press.

Rohracher, H. (2001) 'Managing the technological transition to sustainable construction of buildings: a socio-technical perspective', *Technology Analysis and Strategic Management*, 13(1): 137–50.

Rowe, P.G. (1987) *Design Thinking*. Cambridge, MA: MIT Press.

Ruikar, K. (2004) 'Business process implications of e-commerce in construction organisations', EngD thesis, Loughborough University.

Ruikar, K., Anumba, C.J. and Carrillo, P.M. (2006) 'VERDICT – an e-readiness assessment application for construction companies', *Automation in Construction*, 15(1): 98–110.

Ruikar, K., Anumba, C.J., Carrillo, P.M. and Stevenson, G. (2001) 'E-commerce in construction: barriers and enablers', *Proceedings of the 8th International Conference on Civil and Structural Engineering Computing*, Eisenstadt, Austria, 19–21 September.

Ruikar, K., Sexton, M., Goulding, J. and Abbott, C. (2006) 'Technology strategies for construction firms', *Proceedings of the Joint 2006 CIB W065/W055/W086 International Symposium* (on CD), Italy.

Santorella, G. (2011) *Lean Culture for the Construction Industry: building responsible and committed project teams*. Boca Raton, FL: CRC Press.

Schön, D.A. (1983) *The Reflective Practitioner*. New York: Basic Books.

Sharp, D. (1986) *The Business of Architectural Practice*. London: Collins.

Shelbourn, M., Bouchlaghem, D., Anumba, C.J. and Carrillo, P.C. (2006), *Planning and Implementation of Effective Collaboration within Construction – A Handbook*, PIECC Project, Loughborough University.

Shingo, S., McLoughlin, C. and Epley, T. (2007) *Kaizen and the Art of Creative Thinking: the scientific thinking mechanism*. Bellingham, WA: Enna Enna Products Corporation.

Simon, E.D. (1944) *The Placing and Management of Building Contracts*. London: HMSO.

Sodagar, B. and Fieldson, R. (2008) 'Towards a sustainable construction practice', *Construction Information Quarterly*, 10(3): 101–8.

Tekla (2012) 'Tekla 3D steelwork solution brochure'. Available at: www.comp-engineering.com/ downloads/brochures/XSTEEL/English/Xsteel%203D%20brochure.pdf (accessed 30 May 2012).

Thyssen, M.H. (2011) 'Facilitating value creation and delivery in construction projects: new vistas for design management', PhD thesis, Technical University of Denmark.

Thyssen, M.H., Emmitt, S., Bonke, S. and Christoffersen, A.K. (2010) 'Facilitating client value creation in the conceptual design phase of construction projects – a workshop approach', *Architectural Engineering and Design Management*, 6(1): 18–30.

Tribelsky, E. and Sacks, R. (2011) 'An empirical study of information flows in multidisciplinary civil engineering design teams using lean measures', in Emmitt, S. (ed.), 'Lean design management', *Architectural Engineering and Design Management*, special edn, 7(2): 85–101.

Tuckman, B.W. (1965) 'Development sequences in small groups' *Psychological Bulletin*, 63, 384–399.

Tzortzopoulos, P. and Cooper, R. (2007) 'Design management from a contractor's perspective: the need for clarity', in Emmitt, S. (ed.), 'Aspects of design management', *Architectural Engineering and Design Management*, special edn, 3(1): 17–28.

Various authors 'Curved mirror'. Available at: Wikipedia, http://en.wikipedia.org/wiki/Curved_mirror (accessed September 2008).

Varughese, J., Bouchlaghem, D., Brocklehurst, D. and Sharma, S. (2010) 'People flow modelling in building design', In W. Tizani (ed.) *Computing in Civil and Building Engineering, Proceedings of the International Conference*, 30 June–2 July, Nottingham, UK, Nottingham University Press, Paper 30, p. 59.

Volkema, R.J. and Niederman, F. (1995) 'Organizational meetings. Formats and information requirements', *Small Group Research*, 26: 3–24.

Wade, J.W. (1977) *Architecture, Problems, and Purposes: Architectural Design as a Basic Problem-solving Process*. New York: John Wiley & Sons.

Weisstein, E.W. 'Sphere Packing', available from Mathworld – A Wolfram Web Resource, http://mathworld.wolfram.com/SpherePacking.html (accessed September 2008).

Whyte, J., Bouchlaghem, N., Thorpe, A. and McCaffer R. (2000) *Automation in Construction*', 10: 43–55.

Wilkinson, P. (2005) *Construction Collaboration Technologies: The Extranet Evolution*. London: Taylor & Francis, pp. 6–7.

Womack, J. and Jones, D. (1996) *Lean Thinking: banish waste and create wealth in your corporation*. New York: Simon & Schuster.

Womack, J.P., Jones, D.T. and Roos, D. (1991) *The Machine That Changed the World: the story of lean production*. New York: Harper Business.

World Commission on Environment and Development (1987) *Our Common Future* (The Brundlandt Report). Oxford: Oxford University Press.

Wren, D.A. (1967) 'Interface and interorganizational coordination', *The Academy of Management Journal*, 10(1), 69-81.

Yli-Renko, H., Sapienza, H. and Hay, M. (2001), 'The role of contractual governance flexibility in realising the outcomes of key customer relationships', *Journal of Business Venturing*, 16: 529–555.

INDEX